Informational Society

An economic theory of discovery, invention and innovation

Informational Society

An economic theory of discovery, invention and innovation

Alfred Lorn Norman

KLUWER ACADEMIC PUBLISHERS

Boston/Dordrecht/London

Distributors for North America:
Kluwer Academic Publishers
101 Philip Drive
Assinippi Park
Norwell, Massachusetts 02061 USA

Distributors for all other countries:
Kluwer Academic Publishers Group
Distribution Centre
Post Office Box 322
3300 AH Dordrecht, THE NETHERLANDS

Library of Congress Cataloging-in-Publication Data

Norman, A. L. (Alfred L.)
Informational society : an economic theory of discovery, invention, and innovation / Alfred Lorn Norman.
p. cm.
Includes bibliographical references and index.
ISBN 0-7923-9303-1 (alk. paper)
1. Information technology--Economic aspects. 2. Information technology--Social aspects. 3. Communication--Social aspects. 4. Technology and civilization. I. Title.
HC79.I55N67 1993
303.48'33--dc20 92-34667
CIP

Printed on acid-free paper.

Printed in the United States of America

Contents

Preface

Writing this book has been an odyssey which has occupied over twenty years of my professional career.

The most indelible impression of my early professional career was that society constantly promotes and adjusts to technological change. As the engineer officer of a destroyer escort, I supervised major shipboard modifications to accommodate new military technology. Later as an aerospace engineer, I developed an optimal trajectory program which was used to analyze the performance of the SIVB stage of the rocket which sent man to the moon. My experiences in the military and in the aerospace industry made me keenly aware of the importance of technological change in society.

Deciding that while moon walks were interesting, they would do little to alleviate social problems here on earth, I endeavored to become an economist and use my math and computing skills in development economics. In pursuing my PhD degree I was stunned to find that although earlier economists such as Marx and Schumpeter were deeply interested in technology, the mainstream economics taught by the faculty assumed tastes and technology as given and thus had almost nothing to say about how society promoted and adjusted to technological change. At that point, I decided to pursue my own research agenda.

As an academic, I decided to write the minimum number of technical articles in prestigious journals to gain tenure and devote the majority of my energies to constructing an utopian design for the emerging microbinics technology. The writing of this book turned out to be a gargantuan task. To clearly state what I wanted to say required expanding the scope of the manuscript to cover most aspects of the political economy. And, the more I expanded the manuscript's scope, the more voluminously I had to read to gain an understanding of existing social arrangements. In some cases writing ten pages of text required reading hundreds of pages of material. Not only that but in rare cases where no theories existed upon which to base my utopian design, I had to construct them. Writing the early drafts of the manuscript took about three years each.

With each revision my thinking progressed requiring more reading. For example, at first I envisioned communication technology making government at all levels direct democracy. Later I realized that as government became more and more complicated, the average citizen would have no interest in participating in issues which did not directly affect him or her.

The final draft presents a utopian design for society in approximately the year 2050. The purpose of this design is to promote discovery, invention and innovation which have become the principal aspects of nation state competition. Because the design requires major institutional changes which would require a long period to gain even partial public acceptance, the design had to be is presented in the future. The future design is based on technology coming into existence.

This book is an academic book; nevertheless, it should be of interest to a broader audience than simply academics. The manuscript was used as a text for many years in my upper division course, Informational Society. With each revision I had my best students review a chapter. Any paragraph or sentence which they said was incomprehensible was rewritten. The book should be accessible to intelligent adults who have had some training in undergraduate economics and political science. Technical references are directed towards social science readers; however, as the technology is changing so quickly, they are merely a snapshot in time.

I will consider this book a great success if it convinces its readers that the technology of the next century will create vast social changes which will require major institutional changes. While the reader may not agree with my proposed utopian design, in fact he or she may strongly disagree, the book is a success if at least the reader recognizes that major institutional changes are required to achieve better social performance.

I would like to thank my family, especially my wife, Noreen, for enduring the years of frustration while I wrote this book and went through the rejection process of trying to find a publisher for an unusual book. I also would like to thank my colleague, Dan Slesnick for his encouragement. I thank my editor, Mary Trachsel, for demonstrating to me the standards of English to write a readable book are much higher than the standards of English to string equations together in a publishable technical article. Moreover, I would like to thank the numerous graders in Informational Society who helped me make this book more readable.

In this regard, I am especially grateful to Jenny Womack, Darrin King and John Lubrano. Finally, I would like to thank Dave McConnell, Rose Antonelli and Zachary Rolnik at Kluwer Academic Publishers whose enthusiasm made the final push to organize the manuscript into camera ready format enjoyable.

Chapter 1

Discovery, Invention and Innovation

Introduction

This book proposes major changes in the United States' government. Because these changes would require a long gestation period to gain public acceptance, they are proposed for a forecast of the United States' political economy in the middle of the 21st century. During this period, the United States and other advanced societies will complete the transition from industrial to informational societies. Consequently, the goal of the new governmental design will be to promote the political economy of informational society.

The magnitude of the changes in the political economy in the transition from industrial to informational society will be as great as that of the previous changes which took place in the transition from mercantilist to industrial societies. In this transition the central concern in the study of political economy shifted from trade regulated by mercantilism to production of goods and services regulated by *laissez faire*. For example, Adam Smith in his *Wealth of Nations*[1] proposed first, the elimination of the trade and other economic restrictions proposed by the earlier mercantilists, and second, the promotion of a competitive political economy as the ideal social arrangement to prosper in an industrial society. In the shift from industrial to informational society, the central concern in the study of political economy will shift from production to discovery, invention and innovation.

In the post World War II world, nations must achieve a rapid rate of discovery of new knowledge and a rapid development of the associated inventions and innovations to compete politically and economically.[2] For example, prior to the current slowdown in the arms race, military competition between the superpowers has been a race to apply science to gain a technological advantage such as first strike capability or Star Wars defense[3]. Economic competition has also been directed toward technological supremacy by applying science in product function and production efficiency[4]. For example, the strategic element in the competition for microelectronics markets is the striving by electronics firms for efficient production of ever more powerful memory chips. As the US is learning through competing with the Japanese, competition involves not only inventing a stream of new products, but also innovating a stream of improved procedures to produce ever more reliable products at lower costs.

Most attention of social observers is focused on the immediate problem of US manufacturing innovation[5]. While US research universities are productive and US firms inventive in creating new products, US manufacturing firms lag in innovations in manufacturing in such areas as quality control and robotics. Numerous remedies have been suggested including placing greater emphasis on manufacturing in lieu of finance, educating a more technologically competent workforce, and placing much greater emphasis on long time results in decision-making. The author assumes that over the next decade or so many of these suggestions will be adopted and the US manufacturing will regain an internationally competitive rate of innovation in manufacturing.

However, attention solely to manufacturing innovation is far to narrow to make the US more competitive. This is because a national political economy which wishes to maintain an internationally competitive position must also generate a correspondingly rapid rate of governmental and individual innovation to advantageously cope with the rapid rate of private invention and innovation. Government must innovate to cope with a rapid rate of technological change which embodies both new opportunities for increased governmental efficiency and new problems which frequently will require new regulation. For example, federal, state, and local governments must innovate to apply advances in office automation technology to reduce administrative costs. Also, new technologies frequently create new hazards with which government must innovate a socially efficient regulatory mechanism.

In addition, the successful incorporation of new technology into society requires innovation by individuals and groups in a far broader scope than just manufacturing innovation. Individuals will not use new technology unless it can perform some social function better than older technology. And the development of successful applications of new technology by individuals and groups is a form of innovation. Consequently, innovations are required in all aspects of daily life from business decisions to personal lifestyles. For example, currently individual experiments with telecommuting lifestyles will determine the conditions under which telecommuting is desirable to the individual and simultaneously improves economic efficiency.[6]

The reason for considering government central to the promotion of discovery, invention and innovation is that government through such factors as decentralization, intellectual property law, regulation, research funding and tax incentives creates the institutional framework for discovery, invention and innovation. We shall argue that the 18th century governmental design for a primitive agricultural country is now hopelessly inadequate to cope with the problems of the 21st century discovery, invention and innovation. Major governmental changes will be proposed. But in as much as these changes would require voter approval they would require a long gestation period to be adopted even in modified form. Thus the governmental design is based on a forecast of informational society, the political economy in the mid next century.

A requirement to achieve a good governmental design for informational society is a theory of discovery, invention, and innovation. Unfortunately, mainstream economics is not focused on these activities. Microeconomics[7] for example, considers the impact of price changes on supply and demand for fixed, given technology. And in standard microeconomics the problems of discovery, invention, and innovation are considered external to price theory. Moreover, while numerous investigators have individually proposed discovery, invention, and innovation theories, none of these are widely accepted. To provide a theoretical basis for a revised political economic design, the first step is to present a theory for the design. From this theory it will be apparent the improving the rate of innovation, broadly defined, is an extremely difficult problem.

Definitions: Discovery, Invention, and Innovation

The starting point in developing such a theory is to define terms, which will be indicated by *italics* if they recur throughout the book. The first two definitions to consider are discovery and knowledge. A *discovery* is a new increment to *knowledge*[8] which is defined as understanding the behavior of observable natural phenomena as well as understanding the structure of logical relationships. These observable natural phenomena consist of all physical, biological, and social processes which can be directly or indirectly observed. Logical relationships, on the other hand, are described by the study of pure logic, mathematics, statistics and computer science.

Knowledge of behavior can be further partitioned into theoretical models and empirical relationships. Theorists create models to explain and forecast behavior and empiricists observe behavioral relationships. Their pursuit of knowledge varies from the most theoretical to the most practical, or alternatively, from deep to surface. For example, theorists create models which vary from intuitive qualitative models to formal mathematical models, whose predictive capacity depends on how well the behavior being modeled is understood. Similarly, empiricists make observations using methodology which varies from carefully controlled experiments to casual unstructured observations.

With definitions for discovery and knowledge, the relationship between these two concepts can be clarified. Some examples of discoveries are the creation of a new theory, the observation of a new behavioral relationship or the development of a new skill. Current knowledge is the sum of prior discoveries which are stored either in human memory or on a human record such as printed material. Thus, not all prior discoveries are part of current knowledge because some discoveries are forgotten. Examples of forgotten discoveries are skills associated with obsolete technology such as the techniques for constructing stone-age tools, which modern archaeologists are trying to recreate[9].

The next definition to consider is an invention. An *invention* is a new manmade device or process. A new device which qualifies as an invention may take such forms as a new physical product, a new biological lifeform or a new piece of software. A process, on the other hand, is a chemical, physical, or biological chain of events that produces a product or service. To be patentable, an invention must meet a test of originality. But the fact that an invention may qualify for a patent, does not guarantee that the invention will be profitable to produce. Each year inventors create numerous inventions, of which only a small percent will be profitable to produce. Corporations, in fact, focus much inventive effort on making improvements to existing products and processes. These improvements will be considered minor inventions, regardless of whether or not they could be patented.

Finally, we must define *innovation* which simply put, will be defined as a better way of doing things. Individuals and institutions innovate in all their *goal-directed behavior*[10] which is defined as an effort of an individuals or an institutions at achieving performance as measured by a criterion, whether objective or subjective. An example of goal-directed behavior is the effort of a firm to maximize profits. Another example would be the striving of a politician to obtain re-election. A third example, based on a subjective criterion, is the effort of a consumer to select a purchase most agreeable to his or her tastes. Thus, goal-directed behavior and consequently, innovation encompass a broad range of economic, political and social behavior. With respect to such goal-directed behavior, a formal definition of an *innovation* is the creation or implementation of a new alternative which achieves higher performance as measured by the respective criterion.

Goal-directed behavior provides numerous opportunities for innovation because one aspect of almost all such behavior is selecting between alternatives to achieve the goal. A manager of an existing product line might consider producing the product line at the firm's current plant, contracting the product line to be produced in the Far East, or constructing a new plant which would implement new automation technology. If the manager implements the new automated plant and achieves higher profits, the implementation of this alternative would constitute an innovation.

Generally, an innovation is required in order to make an invention profitable. The invention of spreadsheet software, for instance, became profitable as businessmen innovated a large number of applications for spreadsheet analysis. But, an innovation does not necessarily need to be based on an invention. For example, the development of the assemblyline in manufacturing is an innovation which is based on the organization of production rather than a specific invention.

Since the decision to develop a successful invention by a firm is the selection of an alternative which improves profit-striving performance, inventive activity by firms can be considered a special category of innovation. A point made throughout the book is that invention is quite different from other categories of innovation. For example, property rights for invention differ markedly from property rights for other types on innovation. To keep the terminology as simple as possible, innovation, in all subsequent discussion, will refer to all categories of innovation other than invention.

Interactions and Knowledge Accumulation

Once discovery, invention, and innovation have been defined, the next step is to describe the interactions among these activities. An important empirical fact to be considered is that the interactions between these activities are two way. For example, discoveries frequently lead to inventions, but in many cases the full economic development of an invention requires major new discoveries. Also, invention and innovation can lead to a large increase in knowledge accumulation. Fortunately, because the knowledge required for invention is specialized, a successful inventor has to understand only a portion of this knowledge accumulation.

To illuminate these relationships, let us start with the interactions within the process of discovery itself. An important component of discovery is the direct pursuit of basic knowledge by scientists and mathematicians. Science has become a specialized activity which advances through the interactions of specialists such as empiricists, theorists, mathematicians and engineers. Empiricists discover new phenomena, which stimulate theorists to explain with new models and theories. Theorists then use formal mathematical methods to deduce the implications of their models. Empiricists either confirm or reject these implications on the basis of experiments and hypothesis testing. In addition, engineers using new discoveries create new instruments such as new observation devices and computers, which in turn promote further empirical and theoretical studies. Similarly, mathematicians in discovering the structure of formal relationships, create new tools which theorists can use in constructing formal models of behavior. Through the interactions of specialists, then, new formal models for explaining and predicting behavior are created[11].

The next step is to describe the interactions between discovery and invention. Discovery frequently creates opportunities for invention; however, the development of an invention generally requires further discovery specific to that invention. Inventors can rarely invent a fundamentally new product as a pure exercise in engineering. That is, they can rarely design a product purely from known principles, since theory rarely provides answers to all the design questions which are likely to arise during the process of invention. For example, in designing the airplane, the Wright brothers had to conduct wind tunnel experiments to design an efficient airfoil. They could not simply apply the nascent aerodynamic theory which existed at that time.

The relationship between theoretical discovery and invention is two way. An example of a theoretical advance which led to invention is the case of nuclear power. With Einstein's development of a theory to explain the relationship between mass and energy, physicists and engineers made numerous applied discoveries first to build the atomic bomb and then to create atomic power. In contrast, the advance of the economic usefulness of an invention from satisfying the legal definition for patentability to widespread economic application frequently requires a major investment in theoretical discovery[12]. The development of commercial aviation and military airforces, for instance, stimulated the advance of theoretical discoveries in aerodynamics. In like fashion, the current efforts to develop superconductivity will undoubtedly generate major new theoretical models of superconductivity.

Moreover, discovery to develop the inventions associated with a new technology and its applications frequently leads to a significant accumulation of knowledge in many fields. Consider, for instance, the inventions associated with integrated circuits, computers and software. Crowding more and more components into an integrated circuit advances knowledge of materials at the atomic level, and designing integrated circuits advances knowledge of silicone compilers, software used in the design process. Reducing the production cost of computers requires discoveries to create new production techniques such as surface mount technology. One aspect of software discovery are advances in artificial intelligence to make programmers more efficient. The development of new applications software is frequently based on discoveries in the respective applications discipline. For example, the development of the mathematics of linear programming led to the development of software for applications.

If inventors in an industry had to understand all the knowledge accumulated in the development of technology in that industry, the process of invention would be severely hampered by the knowledge requirements. But, because technology generally has a modular structure, the process of invention requires specialized knowledge and discovery. Consider, for instance, the new central processor unit, the 80486 chip, created by Intel. An inventor using this chip to invent a new computer workstation needs to know only the operating characteristics of the 80486 chip and not the knowledge needed to design the chip, crowd the components onto the chip, or obtain economic yield rates. Similarly, the developer of the operating system needs only to know the operating characteristics of the new computer rather than information about how it is designed. Finally, the developers of applications software knowledge of the computer are limited to knowing the software language and the operating system. This modular specialization of technology greatly facilitates invention by limiting the knowledge requirements for invention.

Within the modular structure of technology the depth of knowledge required for effective invention varies greatly. An inventor has an incentive to understand theory to the extent that this knowledge reduces the applied empirical research needed to perfect an invention. Some inventive activity takes place by inventors who understand how the modules work and creatively combine them. One example is the creation of the Apple II personal computer by Steve Wozniack who used a microprocessor designed for intelligent appliances as one of his modules[13]. On the other hand, it is doubtful that a quantum effects transistor could be developed by an inventor without a considerable formal study in quantum mechanics[14].

Next we shall consider the interactions between discovery and innovation. Innovation, like invention, can involve a two-way interaction with theoretical discovery. One example of a theoretical discovery promoting innovation is the development of the theory of statistical quality control. The applications of this theory to production have led to a significant decrease in the number of rejects. Innovation can also promote theoretical discoveries. The series of innovations in the automation of manufacturing is stimulating all types of discoveries in machine intelligence.

Much innovation involves the interaction of invention and surface discovery in the form of the development of new skills. Consider, for instance, the innovations in processing paper in modern offices. The innovation of wordprocessing as a displacement of typing required the development of wordprocessing skills and their dissemination through instruction manuals and training courses. The diffusion of stand alone wordprocessors set the stage for numerous inventions such as storage devices, laser printers, and local area networks to improve the flow of paperwork in the office. In addition, new inventions linking the office equipment together electronically promoted numerous innovations such as electronic filing of documents and electronic mail. New software such as desktop publishing programs generate new skills in creating documents, and these skills in turn become the basis for a round of innovations in the use of desktop publishing in business.

The activities of invention and innovation, then, have a multiplier effect on knowledge accumulation. As was previously mentioned, the drive to obtain the full economic impact of an invention stimulates discovery. In addition, by definition an invention requires an innovation to become profitable. This innovation generally includes a significant increase in practical knowledge about how to produce, use, and repair the new invention. But, as a new technology displaces an old technology some of the practical knowledge of the old technology is lost.

This knowledge multiplier can be very large, as inventions and innovations frequently have a cascading effect. For example, industrialization started with inventions in textile machinery and innovations in textile production. Subsequently, by attracting a large number of workers to a central location, industrialization created an urban workforce. By extension, industrialization generated innovations both in lifestyles and in city government in the sense that it created the need for providing public services such as clean water. Fully realizing the social implications and economic advances of a discovery in basic knowledge, then, generally requires a major increase in practical knowledge.

Improving Performance: Discovery and Invention

A discussion of the factors promoting discovery, invention and innovation is needed in order to evaluate the current political economy and in order to propose improvements. As was discussed in the previous section discovery, invention, and innovation are highly interactive activities. Effective promotion of these activities therefore requires the development of institutions and incentive systems to promote the interactions. Let us start the discussion by focusing on discovery, invention and their interaction. For now, the discussion of discovery is limited to the types of discoveries which promote invention.

The promotion of discovery and invention requires the resolution of a conflict between the free flow of ideas and economic incentives. Discovery is based on prior discovery and is consequently promoted by rapid dissemination of new results which anyone can use without charge. Similarly, rapid dissemination of the complete specifications of new inventions, available to anyone without charge, would promote inventive activity directed towards making improvements in the original invention. Suppose the promotion of discovery and invention were advanced solely by the pursuit of fame, the term "the pursuit of fame" meaning that discovery and invention are usually intrinsically interesting activities which provide the successful researcher or inventor with public recognition. If the pursuit of fame were the sole incentive in promoting discovery and invention, then researchers and inventors would be rewarded for making discoveries and the specifications of new inventions known immediately to all interested parties free of charge.

While the pursuit of fame provides an excellent incentive system for the free flow of ideas, it does not provide an adequate financial incentive to promote discovery and inventive activity. The economic incentive, of course, is the profit from the sale of an invention. But if everyone could freely use the specifications of new inventions, there would be little profit in making large investments to develop a new invention which others could quickly produce. Thus, the pursuit of fame leads to economic incentives which are inadequate for achieving a high rate of discovery and invention.

In market societies intellectual property in the form of patents, trade secrets, and copyright have been developed to create better economic incentives for inventive and artistic endeavor. A patent gives an inventor exclusive use of an invention for a finite period of time. A trade secret enables a firm to sue for damages if an employee reveals economic secrets of the firm, and a copyright provides writers and musicians exclusive use of their creations for a finite period of time. Over time, innovations are needed to adjust these intellectual property rights to fit the changing needs of the political economy[15]. For example, in recent years copyright protection has been extended to software and integrated circuit masks.

The creation of intellectual property rights, however, conflicts with the free flow of ideas. Inventive activity, whether protected by patents, trade secrets or copyrights, generally requires directed discovery. As was pointed out, the full economic development of an invention generally promotes considerable discovery. Private funding for the profit motive provides managers no incentive for allowing researchers to disseminate results which might be of benefit to rivals. Accordingly, while property rights promote economic incentives for discovery and invention, they also tend to inhibit discovery by checking the free flow of ideas.

Promoting a high rate of discovery and invention requires the development of an efficient set of intellectual property rights[16]. Historically, no property rights were created for discovery in the form of theory, theorems or ideas. Generally the more theoretical the discovery, the less obvious the potential economic applications. For example, George Boole, the 19th-century mathematician, created Boolean algebra as an intellectual exercise. But not until a century later has Boolean algebra become an important tool for the analysis and design of electronic circuits. Discoveries such as Boole's, because their economic applications are not immediately obvious, are unlikely to be promoted by property rights. In fact, the attempt to create property rights for discovery would probably reduce the rate of discovery by greatly inhibiting the free flow of ideas.

The development of efficient intellectual property rights in order to promote invention must also take into account the adverse economic incentives created for rivals of the one who holds intellectual property rights. Patents create incentives for firms to perform research that will enable them to circumvent the patents of their rivals. While this research effort may increase the profits of the firm, it is less likely to produce a significant new invention. Such behavior is discouraged by laws and legal precedents that make the patent infringement suits more likely to collect damages. If, on the other hand, the laws and legal precedents make the successful pursuit of patent infringement suits too easy, large firms will use the legal costs of such suits as a competitive weapon to intimidate small inventive firms.

Another example of adverse incentives of intellectual property is the fact that trade secrets encourage firms to reverse engineer other firms' products. For example, before integrated circuit masks were protected by copyright, rival firms would construct the designs for other firms' integrated circuits from the product. While changes in the law modify economic incentives, it is, for the most part, the courts that determine the economic incentives of intellectual property law by setting precedents in cases. For example, software copyright provides a software inventor property rights to "the look and feel" of a copyrighted program. How this ambiguous phrase is defined by court precedents will determine the economic incentives created by this right. Because discovery must necessarily precede invention, efficient intellectual property rights indirectly create economic incentives for discovery. In considering the finance of discovery, then, the first question is whether firms and individual inventors will devote sufficient resources to promote discovery. As the economic value of an invention is difficult to predict, the economic value of investment in discovery which might lead to invention is even more difficult to predict[17]. Without analytical techniques for predicting the profit from an investment in discovery, businessmen determine the amount of such investment by rules of thumb learned over time. Through competition and imitation, the amount of such investment in most industries has gradually risen over time.

Given the cognitive limitations of managers, firms favor applied research wherein the managers can reasonably expect profits within a limited time horizon. The appropriate time horizon at which efforts at discovery can be expected to yield profits has sparked a fierce debate. Because American managers are pressured by pension funds, their bottom line mentality tends result in rules of thumb favoring immediate profits at the expense of long run competitiveness[18]. In this regard Japanese firms are reputed to have better rules of thumb than their American counterparts.

Regardless of how this debate is settled, firms are unlikely to engage in enough basic research to promote long term invention. First, because the results of basic research are in the public domain, firms have no mechanism for obtaining the economic value of their efforts other than beating their rivals in a race to create the implied inventions. Second, the profitable implications of basic research are often very difficult to forecast; and third, the time horizon for obtaining profits from basic research can be long.

Government investment is therefore essential to the funding of basic research which is not likely to yield a payoff within a moderately short time horizon. The public funding of basic research promoting pure discovery tends to partition discovery and invention into two diverse cultures. Publicly funded basic research generally takes place in research laboratories where the incentive system is the pursuit of fame, and thus promotes the free flow of ideas. The private promotion of discovery leading to invention, on the other hand, is motivated by potential profits from inventions and generally takes place in an atmosphere of secrecy.

The discrepancy between incentive systems for these two types of knowledge accumulation creates two important problems. First, the vast difference between the incentives for basic research and those for invention creates a serious gap in knowledge accumulation. The incentive structure for basic research is based on the production of original, publishable results, whereas the incentive structure for invention is based on promoting those discoveries for which potential economic applications can be reasonably predicted. Between these two lies a gray area of research which is not prestigious enough to be considered a topic for basic research and which lacks the immediately perceivable economic payoff which is required to promote inventive discovery. An example of this gap can be seen in the development of nuclear power, which required the motivation of World War II to prompt the funding of enough applied research to first harness nuclear power to destructive purposes.

Second, the gap between basic research and market cultures does not create strong incentives for the transfer of ideas from pursuit of fame researchers to pursuit of profits promoters. The pursuit of fame promotes communication among other researchers and not between researchers and promoters of market applications. The promotion of a rapid rate of discovery and invention requires specific consideration of this transfer problem.

Because of the gap between the pursuit of fame and the pursuit of profit, government has incentives to fund some applied research as well as basic research. The creation of government support for research raises many difficult questions. A political mechanism is needed to select between competing research projects. This means that the research award mechanism must be able to decide not only between competing research projects, but also decide the boundary between applied research which should be funded by government and the applied research which should be funded by private parties. Because this boundary is fuzzy, government funding of applied research creates incentives for private parties to manipulate government into funding projects which otherwise would have been funded by the private parties themselves. Finally, public funding of applied research raises the question of who should own the property rights to the resulting inventive activity.

Having considered the factors which promote discovery and innovation, let us now consider the factors which promote innovation. In the next two sections we shall provide a more detailed definition of innovation and then discuss the factors which promote innovation. As we shall see the problems of promoting innovation are more difficult to solve than the problems of promoting invention.

Bounded Rational Model of Innovation

A discussion of the factors which promote innovation by individuals or institutions requires a more detailed description of innovation. Because innovation has been defined as the selection or creation of a new alternative which leads to improved performance in a goal directed behavior, such a description requires a model of goal-directed behavior. In this book the model of goal-directed behavior will combine aspects from bounded rationality and cognitive decision theories[19]. Also, the model will be based on the empirical fact that decision-makers expend resources in making decisions.

A basic tenet of these theories of behavior is that humans have limited cognitive skills. While the advance in computer technology has initiated an increase in the analytic component of decision making, most aspects of purposeful behavior are still intuitive. Intuition, moreover, is largely a product of the individual's personal history. Individuals perceive selectively, based on prior experience, and given their limited ability to integrate facts, they are sequential problem solvers. To solve a complex problem, then, they must simplify the problem by using heuristics, that is generalizations which enable them to construct a simple model of behavior. In constructing such a model which accounts for prior experience, individuals must mentally reconstruct prior events.

Because of these limitations, individuals approach purposeful behavior as a hierarchy of tasks to be performed in sequence. For example, a consumer performs the weekly task of grocery shopping by driving to the grocery store, selecting the items sequentially one by one, checking out, and driving home. Such tasks may occur at regular or irregular intervals. An example of the latter would be career choice decisions.

All such tasks, however, whether undertaken at regular or irregular intervals, are assumed to consist of five steps, each of which are affected by man's cognitive limitations. The first is to generate the objective; the second is to determine the alternatives; the third is to evaluate these alternatives; the fourth is to select from the alternatives; and the fifth is to execute the selected alternative. In this work, the first four steps are considered the choice or decision problem. The last step, however, requires action to achieve the goal. The delineation of task behavior into five steps is not meant to imply that these steps are executed in a rigid, sequential order. For example, consumers recursively selecting items for their shopping baskets in a grocery store tend to combine steps.

Because of cognitive limitations, individual task performance will generally be less than optimal. Indeed, faced with a complex task involving a large number of alternatives, *bounded rational* individuals will seek the satisficing rather than the optimal alternative[20]. Simply put, individuals will seek a good, but not necessarily the best, alternative.

Numerous factors limit the performance of satisficing behavior. Psychological research has catalogued numerous defects in individuals' intuitive judgment and choice processes[21]. With respect to the five step task model, judgment is required in the third step: evaluating the alternatives. For most tasks individuals must forecast the future behavioral consequences of selecting an alternative. These forecasts are generally intuitive judgments or estimates. Human judgment, however, is dependent upon the order in which data are presented. Confronted with apparently complete data presentations, humans will tend to overlook critical data omissions. Similarly, too much data can overload human mental capacities, thereby reducing consistency of judgment. Also, humans are poor at making nonlinear forecasts such as predicting the social implications of the advances in computation.

Moreover, most judgment situations involve uncertain phenomena for which man lacks the neurological circuits for optimal intuitive processing. One of the many defects in handling uncertain phenomena, for instance, is that man tends to assume erroneously that the characteristics of a small sample are characteristic of the general population. Furthermore, in combining uncertain data, humans tend to favor the concrete over the abstract. Also, in judgment situations where they should combine new data with old data, humans will frequently ignore the old data. When they do combine the two, they are generally more conservative than optimal.

To illustrate the problems individuals have in combining uncertain data, consider the problem of forming a judgment concerning the reliability of a new automobile. If a friend has recently bought a model which turned out to be a lemon, this concrete data will tend to be given much more weight than a previously conducted study indicating that the model, on average, is reliable. The individual may, in fact, ignore the statistical study altogether. To illustrate the conservative nature of data on judgment consider the forecast that in the 90's US automobiles should become as reliable as the Japanese automobiles. But because Americans are likely to rely more on experience than statistical studies, there will be a time lag before the public acquires the general perception that US automobiles are as reliable as the Japanese.

In addition to affecting human judgment, cognitive limitations also affect the ability of individuals to make choices. Two aspects of making choices in the task model are generating the objective and selecting among alternatives, steps one and four. Individuals must constantly adjust the objectives of their tasks to new conditions in a political economy with a rapid rate of discovery, invention, and innovation. The basic assumption underlying these adjustments is that the objectives generated by individuals are in fact, only estimates of the true objectives. Moreover, experimental evidence demonstrates that individuals do not form consistent objectives[22] and tend to confound their beliefs about behavior with their preferences or tastes. In making choices man uses rules of thumb– simple common sense rules to select alternatives–to reduce the mental effort.

Individual task performance is also limited by the costs of performing the various steps. For example, determining the attributes of products in the marketplace can be vary expensive. Recently, it became public knowledge that some brands of latex paint contain mercury and arsenic. Yet because of the cost involved, individual consumers almost never consider performing chemical analysis of the products they buy, assuming, sometimes erroneously, that they will be protected by government agencies or the product liability laws.

Because of the costs of decision-making and the cognitive limitations of humans, individual satisficing performance can be prone to serious errors. For example, humans can make make errors in judgment in predicting future behavior involving such factors as reliability and safety. These errors can have dire consequences. For example, consumers who bought three wheeled recreational vehicles have tragically lost children in accidents when these vehicles turned over.

The model of goal-directed behavior which we have now developed for individuals will also be applied to institutions. For the purposes of this book, an institution is a group of individuals with a common goal organized in a computer-communications network and bound together by an incentive system. Because of cognitive limitations, institutions also approach goal-directed behavior as a sequence of tasks, which can be broken down into the five steps listed for individuals. Although institutions generally expend much greater resources in decision-making than individuals do, institutions are also prone to serious errors in satisficing performance. Witness, for example, IBM's disastrous introduction of the PCjr to the home computer market.

Because both individuals and institutions are prone to making serious errors in decision-making, what constitutes an innovation is far from obvious. Innovation is not synonymous with the selection of a new alternative. To determine what constitutes an innovation requires a procedure for judging whether the selection of a new alternative improves performance. The problem is to determine the objectives for which performance is to be measured. Because individuals, firms and the government face constant technological change, their respective generated objectives are assumed to be only estimates of their true objectives. This creates a problem because innovation should be defined as the selection or creation of a new alternative which improves the performance of a goal-directed task as measured against the true, but unknown objective.

Let us start to solve this problem by considering the properties of the true, but unknown objectives. In this book it is assumed that the true, but unknown objective of a firm should be to maximize profits. The true, but unknown objective of government will be given the archaic term, "to promote the *common weal*", in order to distinguish the concept from contemporary social welfare theory based on utility theory[23]. The task of government to promote the common weal involves the responsibility to deal with external threats, maintain internal order, promote prosperity and ensure the quality of life. No agreement, however, exists in identifying the basic elements of the common weal or in estimating the tradeoffs between them. In the case of the individual even less can be stipulated. It can only be said, at this point, that the true, but unknown objective of individuals is not considered to be equivalent to the desires of the individual.

Because true objectives can not be precisely defined, criteria are needed by which to judge whether the creation and implementation of a new alternative by an individual, firm, or government is an innovation. For the firm, the selection of a new alternative can be considered an innovation if the change increases profits. In a competitive market, then changes which persist and are imitated can be considered innovations. Similarly, for activities in which governments compete such as efforts to attract investment in their jurisdictions, change will be considered an innovation if the change persists and is imitated.

Not all criteria for judging whether change is an innovation are based on competition. In general, the selection of a new alternative by government can be considered an innovation if the new alternative has the properties of consistency, general benefits, and efficiency. It will be argued in Chapter 2 and 8 that the extent to which these properties can be achieved depends on the system of checks and balances. Change in individual tasks can be considered an innovation if it persists, is imitated and does not diminish the common weal. The final requirement of not diminishing the common weal is needed to eliminate pathological behavior.

Now having defined criteria for determining what is an innovation, let us consider some specific cases. Examples of innovation by firms in the task of production are the development of interchangeable parts in the 19th century and more recently statistical quality control. The creation of an organizational structure with fourteen levels of management at General Motors as compared with five at Toyota would not, however, be considered an innovation[24]. In the realm of government, an example of innovation in the task of distributing social security checks is direct deposit into the bank accounts of the recipients for the sake of convenience and security. A new tax loophole, however, would not be considered an innovation since such a loophole would lack general benefits. An example of individual innovation in the task of working is telecommuting by professionals, whereby professionals perform their work from their homes or local offices through a computer communications network instead of having to travel to the central workplace[25]. The emergence of a lifestyle based on smoking crack would not be considered an innovation, however, as the common weal is diminished by the loss of productive capability and the increased likelihood of crime and child neglect.

Because tasks have a hierarchical structure, innovation can also take place in the decision-making process and in the execution of a task. The model of goal-directed behavior implies that each of the steps in a task can, in itself, be considered a subordinate task. For example, as a consumer can expend considerable time and resources in determining his alternatives, this step can be considered a task subordinate to the task of consumption. Thus, an innovation in the task of consumption can be a new, superior way to determine alternatives. In general, innovations in the first four steps of a task can be classified as innovations in the decision process, and innovations in the fifth step can be described as an innovation in the execution of a task. A recent example of innovation in the decision process in the task of maintaining a financial assets portfolio is the use of personal computers together with information services, such as the Dow Jones Information Service, to identify, evaluate, and buy and sell alternative financial assets.

Innovation in the decision-making process can also occur in the first step–generating the objective. For example, the framers of the constitution incorporated checks and balances into the governmental design in order to achieve a good estimate of the common weal. Subsequently, in the 20th century the political reforms, such as direct election of senators, campaign finance laws, and the principle of one-man one-vote, have sought to make the political process more representative. These changes have had a profound effect of the estimate of the common weal. Many social observers consider these changes innovations[26].

Innovations which improve all aspects of task performance are frequently new forms of organization and incentive systems. The original form of the manufacturing corporation was called a functional organization. Coordination of the firm took place between the president and the vice presidents in charge of manufacturing, finance and other functions of the firm. Managers below the level of the vice presidents were specialists such as accounting managers. As manufacturing firms increasingly began to manufacture multiple products this form of organization exhibited a growing defect–that managers were more sharply focused on their functional specialties than on the profits of the various products and product lines. The resulting shift from functional organization to profit-oriented divisional organization in the 1920's was a change which more sharply focused managerial attention on the profits of each product by coupling managerial rewards to the profits of their respective products.[27]. Since profit-oriented divisional organization has persisted and has been widely imitated, this change is an innovation.

Improving Performance: Innovation

The factors promoting innovation are not intellectual property rights but rather steps which will improve innovation-related discovery. One factor is a socially adequate level of funding for innovation-related research. A second factor is an appropriate resolution of the conflict between observations for research and concerns for privacy, proprietary rights and secrecy. A third factor is providing sufficient variation in goal-directed behavior to be able to study the performance of alternatives. A final factor is improving the strategy for implementing an innovation.

Before considering the four factors promoting innovation, let us explain why it would be difficult to promote innovation by intellectual property rights and discuss the implications of this difficulty. Currently, invention is promoted by intellectual property, whereas most aspects of innovation are not. Intellectual property rights for innovation are limited to trade secrets which protect some aspects of innovation in a firm, but as much of this innovation cannot be kept secret, even this protection is limited. Moreover, it is doubtful that an intellectual property right in the form of a patent could be created for innovation, since innovation is not a precisely defined object but rather the implementation of a concept by an individual, firm, or government. Frequently, each implementation of the concept is specific to the implementor. Moreover, advances in innovation are generally incremental. This means that few innovations would have such clear definition and originality to merit the equivalent of a patent.

The fact that most innovation is not protected by property rights means that the motivation for innovation is to improve the performance of a goal-directed task itself rather than to sell innovation as a product on the market. Because innovation lacks intellectual property rights, any success by an innovator will be rapidly imitated[28] in a competitive market. In addition, for many types of innovation, the innovator has positive incentives to promote imitation. For example, innovations by government are imitated by other governments at the same level, and officials responsible for the innovation enhance their reputations by providing the details to interested imitators. A business innovator can obtain financial rewards for promoting imitation of an innovation by establishing a consulting service or by writing a how-to book.

Also, because innovation lacks intellectual property rights, businessmen have mixed motives concerning the dissemination of innovation. Firms generally prefer to maintain secrecy about innovations which give them an advantage over their rivals, unless, of course, there are considerable profits to be made from creating a consulting service to sell their expertise in creating the innovation. Firms also have a great incentive to disseminate knowledge about innovations based on inventions they are selling, and the extension of copyrights to software provides similar incentives for software firms to disseminate information about innovations based on their software.

While the promotion of innovation through new intellectual property rights may not be feasible, the promotion of innovation indirectly by promoting discovery leading to innovations is both feasible and desirable. The first factor to consider which will promote innovation indirectly is adequate funding for innovation-related discovery. Innovation-related research requires public funding for the same reasons that were discussed for invention-related research. As they lack intellectual property incentive, business innovators have few incentives to invest in making discoveries leading to innovations which will be immediately imitated.

Adequate funding for discoveries in the social sciences, the business disciplines, and the industrial and systems engineering disciplines can promote innovations in two ways. First, discoveries can improve the ability of decision-makers to forecast the consequences of selecting new, untried alternatives. Thus, decision-makers striving to innovate are likely to make fewer mistakes. Second, new discoveries in these areas result in innovations in many areas such as incentives, decision-making, organization, and production.

A second factor which will promote innovation-related discovery is increasing the extent to which investigators can make detailed observations of the phenomenon they are studying. Empirical science requires the researcher, at the very least, to be able to observe the respective phenomena by obtaining a representative sample. Today, however, the ability to obtain representative samples is greatly restricted by proprietary rights, government secrecy, and personal privacy. For example, for many studies of firms, the only data available is aggregate data which masks the behavior of individual firms and subunits within the firms. Innovation-related discovery would be promoted by the creation of a scientific information policy which balances the need for observations for empirical science against the needs for privacy, proprietary rights or secrecy.

But a carefully crafted information policy alone is insufficient to provide useful observations of behavior. In most empirical studies of the political economy, the researcher does not perform experiments but rather observes the consequences of the alternatives selected by political-economic agents. For example, a researcher studying the profitability of mergers can observe the impact of mergers on overall corporate profits. With this type of data, sufficient variation in the data to test hypotheses depends on agents changing policies over time or variation in the policies of agents in similar situations. Such data may contain insufficient variation to test hypotheses as, for example, all firms in an industry may adopt the same alternative in a task or the national government may maintain the same policy over an extended period of time.

A third factor which promotes innovation-related discovery is the promotion of variation in public and private goal-directed behavior. Simply put, if all firms (individuals or governments at the same level) adopt the same policy in a task, nothing empirically will be learned about the possible performance of other alternatives. One approach to increase empirical learning is the promotion of political economic experiments by individuals, firms, and government. Such experiments have been performed since World War II. Another approach in political behavior is to decentralize government policies. Because lower levels of government vary in their political philosophies, decentralization would generally induce more variation than a single policy at a higher level.

The final factor to consider which promotes innovation-related discovery is improving the strategy for implementing an innovation. Because the theory rarely enables an innovator to accurately predict the consequences of the proposed new alternative, an innovator must improvise in creating his innovation. For example, a manufacturer installing a new production process frequently has to improvise for several years before the new plant becomes profitable. The successful implementation of an innovation generally requires much applied discovery if the performance of the new alternative is to improve sufficiently to be classified as an innovation.

Generally, discovery for innovation takes place in a very different learning environment than discovery for invention. In most invention situations, inventing is a separate task in which the inventor can usually conduct experiments to perfect a design. In contrast, the cost of experimentation in innovation is so great that the innovator can generally try only a single alternative. Moreover, applied discovery to perfect an innovation is rarely a separate task, but rather this applied discovery is subordinate to the goal-seeking behavior.

Consider, for instance, the innovation in automation in manufacturing. In the 1980s General Motors built a sequence of new and renovated plants such as Hamtramck, Saginaw Vanguard, and Saturn. Each plant was a pilot project for a new technological advance such as MAP, machine automation protocol, a specification to link machines in manufacturing[29]. Furthermore, these plants have been operated to produce products for sale not simply as laboratories to test automation advances.

The fact that much innovation is based upon a single alternative without experimentation creates a problem: how can the innovator perform the applied discovery necessary to perfect the implementation of the new alternative? What makes applied discovery possible at all is that individuals, firms, and government have much greater incentives to innovate in repetitious tasks than unique tasks. Consequently, applied discovery takes place each time the new alternative is employed to perform the task. This basic strategy for innovation used by most innovators will be defined an *improvisatory* strategy.

With this strategy the innovator uses this acquired knowledge to revise the new alternative to improve the performance. Without the constant stream of revisions the performance might not reach the level of efficiency necessary for it to be classified as an innovation. Indeed, in the GM example above, GM seriously underestimated the amount of applied discovery needed to make the advances in automation an innovation and scaled back the technological advances implemented in the Saturn plant.

From the perspective of the previously discussed task model of goal-directed behavior, an innovator can make applied discoveries in all aspects of the hierarchy of tasks subordinate to the new alternative. For a concrete example, consider the implementation of a new, more automated plant to assemble a product. The process of achieving improved performance through discovery and revision is variously called the experience curve or learning by doing. Specifically, performance improves as workers sharpen their skills at performing their tasks in the new plant. Design defects are discovered and corrected. Bottlenecks, for example, in the delivery of parts are discovered and remedied. In Japan, this learning by doing has been institutionalized into quality circles.

While individuals do indeed improve their performance by learning by doing and imitation, man is not an optimal learner[30]. Besides his previously mentioned limitations in combining data, man tends to make spurious correlations and to seek data to confirm hypotheses rather than to test them. Given the limitations of man's cognitive capabilities, the prospect of achieving near-optimal performance would appear slight. For repetitive production tasks with a given technology under competitive conditions, however, man appears to have the learning capacity to approximate optimal behavior, given sufficient time[31].

The rate at which performance in a repeated task improves depends on the strategy for implementing an innovation. With the improvisatory strategy most learning takes place through experience or through the need to solve obvious problems. Innovators perceive little need for systematic variation of variables to obtain empirical discoveries that improve performance. Performance can be improved in many cases by the adoption of an innovation-implementation strategy which incorporates a more systematic approach to learning.

To illustrate the limitations of the improvisatory strategy, we may suppose an innovator uses a cutoff, such as a minimum score on a test or a minimum grade point average as a condition of employment at his plant. As a consequence, the innovator will never observe the performance of potential employees below that cutoff and hence cannot draw any conclusions concerning the validity of the cutoff[32]. With an improvisatory strategy, this cutoff remains fixed until the innovator perceives the cutoff to be a problem. With no policy for systematic variation of such a variable, the innovator cannot develop a systematic innovation-implementation strategy to improve the performance of the chosen alternative.

Another factor which influences the rate of applied discovery in implementing an innovation is the number of imitators involved. With competition any new alternative which produces much better results than current practice will be imitated immediately. Because imitators are likely to evaluate the factors contributing to the success of an innovation differently, imitators are likely to make numerous modifications in imitating an innovator. The more competing imitators generate greater variation, the more empirical knowledge is gained contributing to refinements to a successful innovation.

The creation of a more systematic approach to learning for innovation is far from simple. With a single implementation of an alternative, most aspects of innovation do not constitute a large enough sample for a statistical design. Moreover, variation through imitation is not controlled, and variation in variables over time may require too long a time horizon to be considered worth the effort. Finally, mathematical models of learning subordinate to performance are very difficult to solve.[33]. This means that the innovator mentioned above has no simple way of determining how hiring a sample of workers below the cutoff will improve the performance of his chosen alternative. Because systematic variation presents so many problems, it is assumed in this book that generally the improvisatory strategy, even with imitation, introduces far too little variation for optimal innovation performance.

Nevertheless, the strategy for implementing an innovation can be improved in some circumstances. In tasks common to a large number of individuals or institutions with similar goals, there are economies of scale[34] in specializing the learning activity. This strategy will be called a *separation* strategy. In other words, learning for innovation, like invention, becomes a separate task. Where learning is a separate activity, performance in learning can be achieved through improved methodologies such as statistical procedures and experimental designs. Once the performance of an innovation has been established through experimentation, the innovation will be adopted to improve performance. This strategy is used in agriculture in advanced nations, where research stations experiment with new types of seeds and production techniques and transmit the results to the farmers.

An important factor which will promote applied discovery to improve innovation performance is to shift from an improvisatory strategy to a separation strategy in those tasks for which there are a large number of participants. To adopt such a separation strategy, private firms must create an institutional arrangement, such as a consortium, to perform the research as a separate task. For governmental innovation the ability to implement such a strategy is dependent on how far government tasks are decentralized to lower levels of government.

Commensurate Rate of Innovation

Because of difficulties in evaluating new alternatives and in developing a good innovation-implementation strategy, the promotion of innovation is more difficult than promoting invention. Nevertheless, having a rate of innovation commensurate with the rate of invention promotes the common weal. As was pointed out, an innovation is required for an invention to become profitable. However, a commensurate rate of innovation is more than the minimum rate needed to make new technology profitable. It is the rate which promotes international competitiveness, balances public and private interests and promotes a positive adaptation of new technology into lifestyles.

To promote the international competitiveness of private firms, firms need to innovate in all their activities, not such in manufacturing production. For example, currently new hardware and software in computing and communication provide firms opportunities to innovate new procedures and organizations to more rapidly react to changing market conditions with less cost. One specific example here is the the substitution of teleconferencing for business travel. International competitiveness depends on firms developing a rate of innovation to rapidly exploit all the new opportunities made possible by new inventions. Also, international competitiveness requires government innovation in those services promoting business competitiveness. For example, government must constantly innovate to educate an internationally competitive workforce. With a rapid rate of private invention, government needs a commensurate rate of innovation to absorb the new technology into government services.

In addition, with a high rate of private invention, government must make a correspondingly high rate of public innovation to balance public interests against private interests. For example, private incentives to produce and use DDT spread its use quickly over the globe. Increasing concentrations of DDT in the higher levels of the food chain created major ecological damage such as depleting bird populations prior to public recognition of the problem and the resulting governmental efforts to cope. Currently advances in computation and communication are advancing the explosion of private data bases on households to make private market decisions. While such data bases do improve the efficiency of the market, they create a heretofore unforeseen and not well understood risk to individual privacy. Judges and legislators are struggling to balance private interests for ever increasing household data against a public need for privacy. A rapid rate of public innovation is required to expeditiously balance public and private interests when there is a rapid rate of private invention.

Finally, individuals absorb new technology into their daily lives. For example, child care frequently suffers currently when both parents work fixed schedules in offices and factories. Technology such as automation, telecommuting, and teleconferencing offer parents the possibility of much more flexible work organization to promote better child care. In addition, with most women working more child care needs to be provided in the workplace. Innovations in life and work styles based on emerging technology are needed to promote child care given the assumption that most married women are unlikely in the foreseeable future to return to the role of full-time housewife.

Book Organization and Main Themes

One purpose of developing a theory of discovery, invention and innovation is to analyze in Chapter 2 the development of the United States from the 1780s to the present. It will be shown that much progress has been made in accelerating the rate of discovery and invention, but that much less progress has been made in accelerating the rate of innovation. For example, the current information policy inhibits rather than promotes innovation. In addition, the growth of the federal government has lead to poorer quality estimates of the common weal and has decreased the variation in governmental innovation strategy. Thus, the capacity of current government to innovate is deficient.

Correcting the defects in the governmental design in order to produce a higher rate of innovation will require fundamental changes in government. Since a major governmental redesign will take a long time to gain sufficient public support for implementation, the design must be based on the implications of technology which is coming into existence while the proposed design is gaining acceptance. The technology most important to the proposed design involves advances in microbinics, that is microelectronics involved in computation and communication. The current advances in microbinics are discussed in Chapter 3.

The forecast of the implications of advances in microbinics is presented in Chapter 4. One major impact of this technology will be the gradual advance of automation in both manufacturing and services. This advance of automation will gradually reduce the work week and require a new mechanism for maintaining a politically stable income distribution. In addition, the advances in microbinics will create a *social nervous system*, or a communication network capable of manipulating, storing and transmitting all text, data, symbols, voice, and images. This means that, increasingly, activities can be conducted from any terminal connected to the social nervous system. The locus of market activities will accordingly shift from physical locations to this social nervous system. Consequently, the advance of automation and the development of the social nervous system will give individuals increasing freedom of location.

With automation increasing leisure time, individuals will need an institutional arrangement for organizing their leisure activities. Because individuals can meaningfully participate in only a small subset of all possible leisure activities, they must make mutually exclusive choices. Moreover, as individuals in informational society will have increasing freedom of location in making such choices, individuals will tend to select communities specializing in the subset of activities which they desire. To promote this selection process, the role of the town in the governmental design is to support the lifestyle of the majority. This will lead to increasing community specialization, which will be discussed in Chapter 5. Two advantages of a system of specialized communities supporting individual lifestyles are a more personal social structure and less energy consumption. To properly select among thousands of smaller specialized communities and among large numbers of alternatives in other types of economic, political and personal decisions, individuals will need detailed data specifying the features of their alternatives. The advances in the social nervous system and terminal organization will promote the creation of an increasing number of decision-support systems for making all types of decisions. These systems will provide large databases and numerous software tools for improving decision making.

The use of decision support systems will affect economic, political and social behavior. The widespread use of economic-decision-support systems is considered in Chapter 6. The development of the social nervous system will promote the development of information service firms providing households with decision-support systems for making household consumer decisions. These systems will make markets more efficient and will create strong incentives for producers to make continual improvements in quality including the reduction in safety hazards. Producers will make even greater use of such systems than they do currently in order to more rapidly respond to changing market conditions. An example of a more rapid response to market conditions will be the greater use of the network form of organization.

The widespread use of decision-support systems will make information policy a heatedly debated public policy issue. Chapter 7 addresses the important question of just what information each decision maker should be able to access for each type of decision. This problem can be considered in the context of the social nervous system which enables all individuals to sit down at their terminals connected in a giant database.

In such a system who should have access to what data? This question must be considered in three parts. First, what data should be available for operational decisions? Second, what data should be available for scientific studies? And third, how should individuals, firms, and government be protected from the misuse of information?

While decision-support systems and an appropriate information policy will make individuals better informed citizens, such support for political decisions is insufficient to achieve a better estimate of the common weal without a major governmental redesign. In Chapters 8 and 9 a new governmental design based on a strengthened concept of checks and balances is proposed. The concept of a judicial review is expanded to the concept of a professional review and major changes are proposed in all three branches of government to create better incentives. In Chapter 9 a decentralization criterion is proposed to achieve a better strategy for implementing governmental innovation.

Finally, the increasing use of decision-support systems to improve personal decision-making will be considered in Chapter 10. The social structure of informational society is such that striving for individual success should promote social progress.

The main theme which recurs in a variety of forms throughout the book is innovation. First is the expansion of decision-support systems for all types of decisions. Second is the proposal for information policy. Third is the effort to shift from an improvisatory strategy to a separation strategy for implementing innovation where ever possible. Fourth is the proposed revision in the system of checks and balances to achieve a better estimate of the common weal. And last is the proposed decentralization criterion to achieve greater decentralization and consequently greater learning in government.

Notes and References 1

1. Smith, Adam, 1776, *An inquiry into the nature and causes of the wealth of nations* (Printed for W. Straham; and T Cadell, London)

2. One of the first to recognize this fact and propose a new role for government was V. Bush in: Bush, V., 1945, *Science, the Endless Frontier*, A report to the President on a Program for Postwar Scientific Research, US Office of Scientific Research and Development. Bush's views have become the accepted wisdom, for example see: Spiegel-Rosing and D. de Solla Price(eds), 1977, *Science, Technology and Society*, (Sage Publications: London)

3. A considerable number of intellectuals believe the application of science to produce greater and greater threats is misdirected. For one such view

compare Boulding's article with the others in: Teich, A. H. and R. Thornton (eds), 1982, *Science, Technology, and the Issues of the Eighties: Policy Outlook*, (Westview Press, Boulder) Also see the *Bulletin of the Atomic Scientists* for numerous articles challenging the wisdom of a technological race for ever more sophisticated weapons.

4. There is a growing literature on this point. To some extent discussions of this point are a call for more funding for a particular type of research. For example see: Feigenbaum, E. A. and P. McCorduck, 1983, *The Fifth Generation: Artificial Intelligence and Japan's Computer Challenge to the world*, (Signet: New York). Amusingly enough, the Japanese achieved only a fraction of the goals set out in the Fifth Generation Computer Project. See: Gross, N, 1992, A Japanese 'Flop' that became a launching pad, Business Week, Jun 8, pp 103

5. For just one of the growing literature on this subject. For a book which should become a classic see: Dertouzos, M., R Lester and R. Solow, 1989, *Made in America: Regaining the Productive Edge*, (The MIT Press: Cambridge)

6. Economists generally define economic efficiency as allocative efficiency which means a Pareto optimal resource allocation. In this book we shall define *economic efficiency* as production efficiency which means there is no way to produce more output with the same inputs. In a political economy with a rapid rate of technological advance the potential for producing more output with the same inputs constantly advances. Entrepreneurs innovate to achieve the advancing potential. Mrs. Shirly created F-International to create software using women telecommuters. As this firm provides its workers with opportunities they would not otherwise have had, F-International promotes economic efficiency as defined. See: Shirly, Mrs. "Steve",1985, F International: A unique approach to computer consulting, *Telematics and Informatics* Vol 2 No. 2, pp 165-168

7. For example, see Varian, Hal R., 1984, *Microeconomic Analysis, Second Edition*, (W. W. Norton & Co.: New York)

8. The purpose of a narrow definition is to keep the discussion focused not to denigrate other types of knowledge such as the humanities . For a discussion of alternative forms of knowledge see: Machlup, F, 1962, *The Production and Distribution of Knowledge in the United States*, (Princeton University Press: Princeton)

9. Toch, Nicholas,1987, The First Technology, *Scientific American*, Apr, pp112-121

10. This definition is a generalization of Schumpeter's definition of innovation. Schumpeter J. A., 1934, *The Theory of Economic Development*, Trans Redvers Opie. (Harvard University Press: Cambridge)

11. For an interesting and controversial discussion of the dynamics of scientific advance see: Kuhn T. S.,1962, *The structure of Scientific Revolutions*, (University of Chicago Press: Chicago)

12. Scherer,F.M.,1984, *Innovation & Growth: Schumpeterian Perspectives*, (The MIT Press: Cambridge)

13. Moritz, Michael, 1984, *The Little Kingdom*, (William Morrow & Com-

pany, Inc: New York)

14. Bate, Robert T., 1988, The Quantum Effect Device: Tomorrow's transistor, *Scientific American*, Mar

15. U.S. Congress, Office of Technology Assessment,1986, *Intellectual Property Rights in as Age of Electronics and Information*, O TA-CIT-302 (Washington, DC: U.S. Government Printing Office, April)

16. For a recent survey of the issues, see: Benko, Robert P.,1987, *Protecting Intellectual Property Rights*, (American Enterprise Institute: Washington) For a theoretical analysis of this problem, see: Nordhaus W. D.,1969, *Invention, Growth, and Welfare: A Theoretical Treatment of Technological Change*, (The MIT Press: Cambridge)

17. U.S. Congress, Office of Technology Assessment, 1986, Research Funding as an Investment: Can we measure the returns?,Science Policy Study Background Report No. 12, (US Government Printing Office: Washington)

18. Port, Otis with R. King and W. Hampton, 1988, How the New Math of Productivity Adds Up, *Business Week*, June 6,pp 103-113

19. One of the first expositions of behavioral man in economics is: Simon, H. A., 1945, *Administrative Behavior*, (The Macmillan Company: New York) For a survey of the economic application of behavior man see H. Simon's Nobel laureate address: Simon, H., 1979, Rational Decision Making in Business Organizations, *The American Economic Review*, Vol 69 No. 4, pp 493-513. 11. Psychologists have made numerous empirical studies concerning the decision capabilities of man. Their work is surveyed in the *Annual Review of Psychology* series, for example see: Slovic, P., B. Fishhoff, and S. Lichtenstein, 1977, Behavioral Decision Theory, *Annual Review of Psychology*, 28:1- 39 and Pitz, G. and N. Sachs, 1984, Judgment and Decision: Theory and Application, *Annual Review of Psychology*, 35:139-163.

20. Simon, H., 1957, *Models of Man*. (John Wiley & Sons,Inc: New York)

21. For a survey see Hogarth, R., 1987, *Judgement and Choice, 2nd Edition*, (John Wiley & Sons: New York)

22. Tversky, A., 1969, Intransitivity of Preferences, *Psychological Review*, Vol 76 No. 1, pp 31-48

23. The social welfare theory approach is to construct a social welfare function based on rational individuals with well defined preference orderings. The fundamental reference in this area is: Arrow, K., 1963, *Social Choice and Individual Values*, Second Edition, (John Wiley & Sons, Inc: New York) In this classic Arrow demonstrates the impossibility of constructing a social welfare function from individual preference orderings where the construction of the welfare function must satisfy small set of restrictions. In this book the problem of the common weal is approached from a very different perspective. First what individuals want, may not be good for them. If an entire society were addicted to heroin, then a social welfare function based on preferences would favor heroin even though such an addition would be fatal in international economic and military competition. In society there is frequently a conflict between desire and knowledge. In this book we shall address the problem of how to obtain

an estimate of the common weal given vast differentials in knowledge among individuals in society. It is assumed that the foundations of the common weal rest not in preference orderings or philosophy but in sociobiology. See : Wilson, E. O., 1975, *Sociobiology: The new synthesis*, (Harvard University Press: Cambridge)

24. Hampton,W.J. and J.R. Norman, 1987, General Motors: What went wrong, *Business Week*, March 16, pp 102-110

25. Telecommuting is not, however, an innovation for office support people who lose fringe benefits and are paid on the basis of piece work.

26. Conservatives usually consider the concept of checks and balances a major innovation. See Buchanan, James M. and Tullock Gordon , 1962, *The Calculus of Consent: Logical Foundation of Constitutional Democracy*, (University of Michigan Press: Ann Arbor). While the move to a more representative government is best expressed in Lincoln's phrase, "Government of the people, by the people and for the people"

27. Chandler, Alfred D.,Jr, 1962, *Structure and Strategy*, (The MIT Press, Cambridge, MA)

28. Indeed, imitation in business has now been given the colorful buzzword, benchmarking, which means searching the world to find the best practice to implement in the firm. To facilitate benchmarking clearing houses such as American Productivity and Quality Center have been established. For example, see: Altany, D., 1992, Benchmarkers Unite: Clearing house provides needed networking opportunities, *Industry Week*, Feb 3, pp 25

29. Poe, Robert, 1988, American Automobile Makers Bet on CIM to defend against Japanese Inroads, *Datamation*, March 1, pp 43-51

30. Hogarth, R., 1987, *Judgement and Choice, 2nd Edition*, (John Wiley & Sons: New York)

31. The problem in providing empirical support for this statement is finding a learning situation where technology changes slowly. Studies of tradition agriculture have indicated that traditional farmer's decisions are approximately optimal. For example, see: Hopper,W. D., 1954 Allocation Efficiency in a Traditional Indian Agriculture, *Journal of Farm Economics*, 47 pp 611-24

32. Einhorn H. J.,1980, Learning from Experience and Suboptimal Rules in Decision Making in Wallsten, T. S.(ed), *Cognitive Processes in Choice and Decision Behavior* (Lawrence Erlbaum Associates, Publishers: Hillsdade NJ)

33. Innovation in the context of a single entrepreneur is mathematically a problem in estimation and control. Such problems are intractable, see: Aoki, M., 1967, *Optimization of Stochastic Systems* (Academic Press: New York). For an economic interpretation of this fact see: Norman, A. and D Shimer, 1993, Risk, Uncertainty and Complexity, *Journal of Economic Dynamics and Control*, forthcoming

34. For the noneconomist an informal definition of *economies of scale* is that it is more efficient to pursue the respective activity at a larger scale of operations. Systematic research by small, independent formers lacks economies of scale because there would be too much duplication of effort.

Chapter 2

The Past

Introduction

The starting point to consider a governmental design for informational society which meets the proposed criterion of promoting a high rate of discovery, invention and innovation is to examine the evolution of the political economy since 1790. In this examination a consideration of the development of incentives and institutional arrangements for discovery, invention and innovation is important in order to forecast future advances. Hence, the new governmental design should focus on making improvements which are unlikely to evolve from the current design.

The examination of our political economy will be divided in three periods: 1780-1789, 1790-1889, and 1890-1989. To establish the governmental design of the Constitution as a reference point, discussion will start with the 1780-1789 period. Because the Progressive period, starting in 1895, marks the beginnings of major modifications in government, the period from 1790-1889 will be considered separately from the period 1890-1989. In order to provide background material and to point out to the reader the magnitude of change over time, the principle features of the political economy will be summarized for each time period.

The fundamental concern of the founding fathers in writing the Constitution was to solve the problem of creating an effective government which had limited powers. As the United States could adapt technology from the more advanced Great Britain, discovery, invention and innovation were very secondary considerations. With the exception of the establishment of a system of copyrights and patents, discovery, invention and innovation were not considered in the Constitution.

In the first hundred years, 1790-1889, while manufacturing became a major economic activity, the government remained consistent with the original design. The central government remained limited, regulation of business was by common law, and there were few public goods. Also, until the end of the first one hundred years, universities did little research, inventors used largely trial and error methodology, and innovators used intuitive improvisatory strategies.

In the second hundred years, 1890-1989, services have become the dominant economic activity. Secondly, the government has progressively deviated from the original design as government has become much larger and much more centralized. Much of this growth has been the provision of an increasing array of public goods. At the same time, regulation has shifted from common law to administrative agencies. Also, in the second hundred years, the rate of discovery has accelerated with the creation of increasing numbers of research universities, which have received public funding for research since World War II. In addition, the creation of corporate laboratories and increasing applied research have led to a more systematic approach to invention.

As the United States political economy has advanced from agriculture to manufacturing and then to services, discovery, invention and innovation have become progressively more important factors in national and international competition. The basic thesis of this chapter is that innovations in incentives and institutions have greatly increased the rate of discovery and invention but have not increased the rate of innovation nearly as much. First, current information policy does not provide observations for innovation-promoting discovery. Second, in most cases there has been no advance in the strategy for implementing innovations. Third, governmental innovation is hindered by the failure of the system of checks and balances to encourage the selection of alternatives with the properties of general benefits, consistency and efficiency. Moreover, the current governmental design is unlikely to eliminate all these defects.

Original Design

In considering our country's original governmental design an obvious fact must be emphasized: the Constitution and the Bill of Rights were written for the political economy of the 1780s, not the 1980s. Therefore, to understand the governmental issues from a perspective of the 1780s, a brief review of the 1780s is in order.

Two important factors shaping the development of the colonial economy were cheap land and expensive transportation[1]. Most Americans, given the availability of cheap land, were members of small farm households. And as these households were linked to the towns by very poor roads, internal land transportation was very expensive. Water transportation was cheaper, but given the existing technology and the long distances between settlements even this transportation was costly. With high transportation costs, most goods were produced in the farming household or locally by artisans. In the early American economy, then, much less activity took place through markets than today. The US, rather than constituting one national economy linked to the world economy, consisted of numerous semi-isolated economies, each connected to one another and the world by water travel.

The significant market activities of the 1780s were commercial farming and trade. Larger farmers, such as the Southern plantation farmers who grew tobacco, rice and indigo as cash crops, were more market oriented. But even the Southern plantations were self sufficient in most staples, with the cash crop sold to England in order to buy luxury goods. In the middle Atlantic states smaller commercial farmers produced livestock and grain for market. Lacking an abundance of good agricultural land, and with plentiful fish off the coast, New England developed fishing, ship building and world trade. But even in New England most people were small farmers.

The political economies of the various states were very different from the national political economy of today. First, almost all businesses were sole proprietorships or partnerships, thus very few corporations existed. In fact, a legislative bill was required in order to obtain a charter for a corporation, an institutional form opposed by agrarians. Second, few of the current government activities existed at that time. For example, at that time social welfare was the province of the extended family not the government. As life expectancy then was much less than today, the burden of the aged on their extended families was slight. Nor was the environment an issue as it is today. Although agricultural practices did result in topsoil erosion, as long as land seemed unlimited no action was deemed necessary.

During the period from the first settlements to the 1780s, economic expansion was primarily the result of the increase in land under cultivation and not the advance of technology. During this period knowledge accumulation was slow; with the exception of funds for exploration, no money was spent on basic research. Harvard was a college educating students to be clergymen, not to be research scientists. Because England was much more industrially advanced than the United States, the most effective means for technological progress was to adapt English knowledge to American conditions rather than to devote major resources to new inventive activity. With a slow rate of technological change, the correspondingly sluggish rate of innovation was not perceived to be a problem. Moreover, with a small knowledge base and no urgent need for inventive or innovative activity, the need for extensive formal education was not apparent. For the most part education was acquired through experience, such as growing up on a farm or serving an apprenticeship. What formal education most people received took place in the home or local community.

In the confederation as is the case currently, both monetary and trade policies were controversial. Then as now, people debated the value of an expansionary monetary policy. State legislatures dominated by agrarians created paper money to produce an expansionary money supply favoring the debtor, and legislated restrictions on the foreclosure of farms for bad debts. Such policies were bitterly opposed by the economic elite, the creditors. Similarly, trade controversies arose because economic policy was the sole province of the sovereign states. Trade policy of the time reflected mercantilism rather than free trade, with each state promoting its industries at the expense of its neighbors.

The most important problems of the 1780s, however, were political rather than economic. The confederation was designed to minimize the transfer of power from the sovereign states to the central government, which consisted of a congress, a weak executive and an adequate court. This government was indirectly responsible to the people through the state governments. Its congress had no power to tax, to raise troops, or to regulate commerce; the individual states retained these powers normally associated with a sovereign nation. Thus the confederation government lacked any means of enforcing its laws in the states. Many perceived it as too weak to either maintain internal order or to promote external security. Shay's Rebellion, an armed tax revolt of Western farmers, accentuated its weakness in maintaining internal order[2].

A growing perception of its weaknesses motivated the Confederation congress to convene a Constitutional Convention in 1787 which exceeded its mandate to create a new government. The Constitution created an effective federal government which was directly responsible to the people through the popular election of congressmen. The issue of sovereignty was resolved by making the federal government supreme, but dividing the acknowledged powers of government between the federal and state governments. The federal government was granted authority for defense, foreign affairs, and foreign and interstate commerce and was given some other minor powers such as control over territories. To fulfill these responsibilities, the federal government had the power to tax, to raise an army and to take whatever action was deemed "necessary and proper" to fulfill its specifically assigned duties.

While creating a stronger central government was a central concern of the framers of the Constitution, promoting discovery, invention and innovation were minor concerns. Nevertheless, as these three activities are central concerns of this book, an examination of how these three activities are addressed in the Constitution is important to establish a base point from which to study their subsequent evolution. Invention-promoting discovery was considered a private matter not to be discussed in the Constitution. In regards to observations necessary for innovation-promoting discovery, the Constitution did establish a nascent information policy. The one active requirement for observations was a census taken every ten years as a basis for representation. Also, implicit in the original design is the basis for political economic observations. To formulate policy a legislator implicitly had the power to investigate political economic conditions. This right stems from the right of the British Parliament to investigate established in the 17th century[3]. As this matter was not an issue in state governments, no mention was made in the Constitution.

While the implicit power of Congress could thus be used to promote discovery to improve policy, the founding fathers' explicit concerns about information policy are more directly expressed in the Bill of Rights. In this document information policy is primarily directed toward protecting individuals from the potential abuse of power by government: the First Amendment denies prior restraint in speech, the Forth requires probable cause for searches of papers, and the Fifth prevents self-incrimination.

Currently, in view of the numerous data bases and the mass of transactions data being analyzed, the lack of privacy alarms many individuals. Yet, in the 1780s the current privacy issue did not exist; much less activity took place through the market, and the keeping of records was so expensive that few records were actually kept. Consequently, the individual did not need a Bill of Rights protection for privacy.

With regards to invention, the Constitution authorized the creation of a federal system of patents and copyrights to provide economic incentives for invention. With regards to private innovation, the Constitution is silent. Private innovation is left to private initiative.

Although the framers of the Constitution did not explicitly consider public innovation as defined in Chapter One, they did create a government which promoted such innovation. The applicable criterion for governmental innovation is the selection of a new alternative which has the properties of general benefits, consistency, and efficiency. First, let us examine how the property of general benefits was defined by the framers of the Constitution. Second, let us consider what institutional arrangements were made to achieve the property of general benefits in the selection of an governmental alternative. And third, let us consider the properties of consistency and efficiency.

To understand how general benefits were defined by the framers of the Constitution, it is necessary to consider the political theory upon which the Constitution is based. The framers of the Constitution were spokesmen for the politically conservative economic elite, the large planters and merchants, and as practical men of affairs they discussed the political issues using conventional wisdom rather than the abstract concepts of theorists[4]. During the 17th and 18th centuries, these theorists, such as Hobbes, Locke and Rousseau[5], developed social contract theory to displace religious arguments as the accepted political basis for government. Social contract theory derives from the question: Under what conditions would a rational man in a state of nature enter a contract to create a government?

In Locke's model, the most widely accepted version of social contract theory, the individual's goal is to enjoy his property–that is his life, liberty, and his estate earned by his labor. According to Locke, a rational man would give up some of his freedom to better enjoy the remainder in an ordered political society. As Madison subsequently pointed out in number 10 of *The Federalist Papers*, the purpose of government is to grant the individual the right to enjoy his property. To the extent that individuals have different abilities to acquire property, then, the purpose of government is to preserve inequity. From the perspective of the framers of the Constitution, the resulting design problem was to limit government's scope in order to allow the individual the right to acquire and enjoy property, but at the same time to create a stronger government in order to reduce external threats and preserve internal order.

Thus, in creating an effective federal government, the framers of the Constitution were concerned that this government not exceed its specified powers. Consequently, a condition that the selection of a governmental alternative have general benefits is that the selection not exceed

Constitutional limits. The main problem in meeting this challenge, according to the framers is that in a representative government, the less successful majority tend to interfere with the property rights of the more successful minority. Restrictions on foreclosure of farms in states controlled by agrarians had made this particular concern concrete rather than abstract to the framers.

One device the framers employed to achieve these desired governmental limitations was the system of checks and balances. Proposed by Montesquieu and made explicit in the construction of several of the state governments, this system was designed to place checks on the tendencies of the three government branches to exceed the prescribed boundaries of their authority. The President was given the power of veto, and the Congress could overturn the veto with a two-thirds vote. The President also could appoint judges who were to be confirmed by the Senate. The most interesting type of check was the concept of judicial review[6] spelled out by Hamilton in the *Federalist Papers* and first asserted by Chief Justice Marshall in the 1803 Supreme Court decision, *Marbury vs. Madison*. This concept gives the court the power to limit the other branches of government to powers expressly granted or implied by the Constitution. Passage of the Bill of Rights placed further limits on the power of the government.

A second device to limit government was the imposition of specific restrictions on the state governments. The states could not exercise powers granted the federal government, such as the power to coin money. In addition, the states were prohibited from interfering with contracts. In strengthening the right of contracts, the framers of the Constitution ensured that state governments would not interfere with the individual's quest to acquire property. As a consequence, the economy would evolve through market mechanisms based on individual property rights.

While the framers of the Constitution had a clearly articulated condition for general benefits, they did not even consider such properties as efficiency and consistency in government. This is not surprising because when government is severely limited in its scope and is a very small component of political economic activity, neither governmental consistency nor efficiency are very important. However, we shall subsequently argue that as the size and scope of government increased, governmental efficiency and consistency become more important.

In summary, two features of the original governmental design which promoted innovation were the system of checks and balances and the restrictions on state legislatures. These design features promoted the selection of new alternatives which would remain within the scope of Constitutional restrictions, a condition of general benefits as envisioned by the framers of the Constitution. At the same time the framers of the Constitution did not consider the conditions of efficiency and consistency worthy of incorporation in small government.

An additional feature of the original government which promoted its innovation was the decentralization of its powers. By defining the powers of the federal government explicitly, the framers meant to leave the implicitly defined residual to the states and the people. Later this division was confirmed by the 10th amendment. Because most government activities were decentralized to the state and local governments, the governmental design meant that initially most governmental innovation would take place below the federal level in an atmosphere of political economic competition and imitation.

First Hundred Years: 1790-1889

In its first hundred years, the United States progressed from a second tier agricultural nation to the most advanced manufacturing nation in the world. By strengthening the concept of property rights, the Constitution promoted the growth of the economy through market mechanisms. The major impediments which were overcome in the transformation were the high cost of transportation and capital, not an inadequate rate of discovery, invention, and innovation. As most of the governmental innovations which were responses to industrialization occurred at lower levels of government and in the courts, few federal-government changes were required.

The first step in analyzing the first hundred years is to present a brief summary of the transformation of the political economy during this period. By 1890 the frontier had disappeared from the continental United States and with the expansion of the United States to the Pacific the population had increased from approximately 4 to 62 million. Because of continual improvements in transportation and communication, the United States changed from a collection of small, semi-isolated, regional economies into a single, national market. This transformation meant that economic activity became increasing specialized. Farming, for instance, became more commercial, with the farmer selling a cash crop to satisfy a larger portion of his needs through the market rather than from home manufacturing. Towns, cities and regions also became progressively specialized in their economic activities. The type of adaptation which created the greatest change in the political economy was the growth of manufacturing, which began as a regional specialization of New England.

The transformation of the economy can be seen in the value added data available starting in 1839[7]. From 1839 to 1889 the gross national product expanded by a factor of about 7.6, whereas manufacturing and construction increased by a factor of 15.1. Manufacturing and construction increased from 31% of the output to 61% of the output during this period.

Manufacturing started with the processing of raw materials, such as lumber, and consumer goods, such as textiles. Over time, however, capital goods, for example machinery and steel, became ever more important. As manufacturing developed, the size of the industrial firm expanded and provided jobs for a growing portion of the population. By 1890 less than half of the workers were employed in agriculture.

Concurrent with the development of markets and manufacturing was the development of financial services and commerce. The first bank was incorporated in 1781 in Philadelphia and by 1890 there were over 8000 American banks with over $8 billion in assets. As more and more goods were produced by factory organization, the volume of goods being shipped longer distances to markets for final consumers increased. Entrepreneurs created new marketing institutions such as jobbers, department stores and catalog sales.

The development of manufacturing also modified social organization as individuals and households moved from farms to the cities. In 1790 about 5% of the population lived in towns and cities with over 2500 inhabitants, but by 1890 about 35% of the population was urban. During the period from 1790 to 1860 New York grew from only 33 thousand to over 1 million inhabitants. Not surprisingly, the fastest growing component of government in the first hundred years was city governments which were needed to provide municipal services.

In addition, the growth of the cities created a new lifestyle. Instead of the entire family working as a production team on the farm, the male became the laborer, the wife the shopper, and the children students. The family, rather than being isolated on the farm, lived in a neighborhood linked to the metropolitan world by newspapers. Because the industrialization which promoted urbanization also made the income distribution less even, these urban neighborhoods became income stratified.

The second step in analyzing the first hundred years is to examine the the transformation of the political economy in greater detail. The most important factor promoting the growth of manufacturing was the improvement in transportation. While a rapidly growing population creates a growing demand for goods which favor mass production, industrial production will not displace handicraft production unless the cost of the former is less than the cost of the latter. In order to meet this requirement industrial production generally must proceed on a much larger scale of operations than handicraft production; hence, to be profitable manufacturing generally requires a much larger market. Whether a market for manufactured products is sufficiently large depends on the cost of transporting raw materials and finished products; the higher the transportation cost, the smaller the area over which a manufacturer can successfully compete with local handicraft producers.

If transportation costs had remained at the levels of 1790, manufacturing would have grown much more slowly if at all. Throughout the nineteenth century, however, transportation costs fell continually because of various improvements in transport systems[8]. The first of these were in water transportation. In the East, the state of New York built the Erie Canal linking New York with the Midwest, and subsequent canal construction linked the Great Lakes to the Mississippi River. In the Mississippi Valley river boats were constantly undergoing improvements. With the introduction of the railroad after 1830, ground transportation also improved dramatically . By the time of the Civil War, the East Coast was linked to Chicago, and after the Civil War the entire nation was spanned by railroads.

Because of these and other improvements in technology and because of increasing competition among transport systems, in most cases freight rates decreased throughout the first one hundred years by over an order of magnitude. For example, railroad freight rates gradually fell by a factor of ten from 1830 to 1890. The year round travel of the railroads permitted a better matching of supply and demand of manufactured goods in remote markets which had formerly relied on water transportation. Railroads are much faster than water travel, and in the North water travel is interrupted in the winter.

Other important factors promoting the growth of manufacturing were improvements in communication and finance. Communications improvements facilitated the control of enterprises spread over great distances. The fall in the cost of transportation and the increase in speed vastly improved the mail service, for instance, and the invention of the telegraph in 1839 created instantaneous communications over long distances. Along with these technological improvements the development of banks, the stock market and other more specialized financial intermediaries reduced the cost of capital by increasing the efficiency of channeling savings from rising incomes into investments. Another mechanism to funnel savings into investment was the adoption of a corporate form of organization first in the railroads and then in manufacturing.

The third step in analyzing the first hundred years is to consider the role of discovery, invention and innovation in the transformation of the political economy. Obviously, this transformation was based on a large number of discoveries, inventions and innovations. Even so, during this period there was little concerted effort to deliberately accelerate the rate of discovery, invention, and innovation.

One factor which was not necessary at the beginning of the first hundred years in order to promote a rapid economic advance was a large investment to promote basic research. As long as manufacturing could rapidly advance through intuitive trial and error experimentation to adapt English technology to American conditions, there was little need to consider creating new ideas that would lead to new inventions.

But as the US caught up to England technologically, the need to increase the rate of discovery gradually became a policy objective. The first step in this direction was a by-product of creating an educated workforce for a manufacturing society. As they industrialised, states developed a system of public education. At the federal level, the Morrill Act in 1862 promoted the creation of public universities. Up until this time the only basic research conducted in this country was exploration such as the Lewis and Clark expedition. But in the latter part of the nineteenth century, the major universities began to emulate the German model of a university by emphasizing research as an important university goal[9]. Moreover, a number of individuals founded new universities, such as John Hopkins, specifically for the purpose of promoting research. Thus, in the first hundred years the creation of public education and the research university laid the foundation for the subsequent acceleration in discovery.

During the early development of manufacturing in the US, most manufactured items were adaptations of British technology rather than developments from new discoveries. Slater, a British mechanic, in conjunction with Brown, a Providence businessman, set up the first textile mill in the United States in 1793, but it wasn't until the 1820s, when improvements were made to the designs copied from the mills around Manchester that textiles became consistently profitable in the United States[10]. Furthermore, the mechanical skills used in developing textile machinery were transferable to the development of machinery for the factory production of other goods[11].

The development of new products in the first hundred years generally required very little resources and very few people. Inventors created most inventions using only an empirical trial-and-error development procedure without extensive applied research. The classic example of this method is the folklore surrounding Edison's development of the electric light bulb. With this methodology, inventors during the first hundred years created a high rate of adaptations, inventions, and improvements simply by intuitively exploiting surface knowledge.

However, the folklore surrounding Edison's achievements is misleading in two aspects. First, as educational levels rose throughout the first hundred years, inventors increasingly had engineering or scientific training. Second, Edison himself, in developing a research and development facility at Menlo Park, was a forerunner to the development of a more systematic strategy for invention in the second hundred years.

Besides invention, the private sector made numerous innovations crucial to the transformation of the political economy. In the United States, labor was more expensive than in England; hence, successful adaptation of technology mandated using less labor than the original English technology. One American contribution to manufacturing in this period was the concept of interchangeable parts in mechanical equipment. This innovation reduced labor both in assembly and in subsequent repair. A second major American innovation in manufacturing was assembly-line production, which increases efficiency of labor by greater specialization. Cyrus McCormick had employed both concepts in the production of reapers by 1850[12].

In addition, private innovators improved the organization of manufacturing firms by creating the manufacturing corporation organized by functions[13]. The corporation was a desirable institution for raising capital, as it limited the liability of the stockholders to the value of the corporate assets. The functions of the corporation, such as production, marketing, and finance were departmentalized, and each department was headed by a vice president managing a specialized staff. The methodology of innovation with respect to business organizations was an intuitive, improvisatory strategy with imitation. The corporate form for manufacturing is, in fact, an adaptation of the corporate form developed to promote railroads, and this organization, in turn, was an adaptation of the joint stock trading companies.

In order to discuss public innovation, one fact must be emphasized. During the first hundred years government was highly decentralized. The promotion and regulation of the economy of each state was the province of the respective state government. Moreover, in each state most regulation of the economy was performed by state courts under common law. Also, most of the problems of coping with urbanization were delegated to municipal government. The fact that government was highly decentralized meant that both public and private innovators used an intuitive, improvisatory strategy with imitation by competitors.

From the beginning, government innovated new tasks in order to promote development. The first such task was the promotion of transportation to spur development faster than market incentives alone could do[14]. As financial markets developed and private firms were able to finance larger projects, government promotion shifted from direct ownership of projects, such as the Erie Canal, to subsidies, such as those paid by various levels of governments to promote railroads.

In addition, state governments had to innovate to provide public goods necessary for industrialization. For example, consider education. With industrialization, urbanization and commercialization of agriculture, the need for better education for all segments of society increased, and as both market incentives and charity funds were insufficient to create mass education, the states, starting with Massachusetts in 1827, passed legislation to create mass primary education at the local level. The public high school was created in the cities and the system of public secondary education expanded rapidly after the Supreme Court of Michigan approved the use of local taxes to support public high schools in 1874[15].

Regulation of business at the state level by the courts also led to innovations in the form of precedents defining business law. In the original design the regulation of business was left to the states and originally most states left such regulation to the common law[16]–that is, the sum of precedents established from earlier decisions. By tradition judges had to follow the precedents of higher courts or suffer the embarrassment of having their decisions overturned. As society has changed, however, judges have been faced with new legal situations which are not covered by precedents. The judicial mechanism for innovation is for a judge to create a new precedent for a new legal situation, using an intuitive, improvisatory strategy with imitation from other jurisdictions. As society changes, then, judges innovating new precedents can create substantial changes in the common law. For example, property originally meant land, and legal precedents had to be created to expand the definition of property to cover physical capital. Similarly, prior to the development of large-scale manufacturing, the doctrine of *caveat emptor*, or buyer beware, placed the burden of testing products on the buyer[17]. But through changes in precedents, continuing up to the present, the burden was shifted to the producer. This is socially advantageous in that it is cheaper for the manufacturer to test products than it is for each consumer to do so.

The brunt of the innovations required to cope with urbanization were left, however, to the municipal governments. These governments had to innovate to provide municipal services such as water, sewage, transportation, lighting, education and welfare[18]. Providing fresh water and removing sewage had a very high rate of social return in reducing disease. As part of their vote-getting activities the bossism form of government provided the first welfare services.

Competition between the cities meant that municipal government had to constantly innovate to provide better services for the growing economy. To create streets capable of handling city traffic flows, for instance, street construction evolved from cobblestones to granite blocks to asphalt, and city lighting developed from lanterns to gas lighting to electric lighting. Many of these services constituted natural monopolies in that large-scale production was required to achieve economies of scale, and more than one producer would have meant expensive duplication of delivery systems. In the search for incentive to further the economic performance of natural monopolies, cities relied either on franchises or city owned production units.

The last aspect of the decentralized government of the first hundred years to consider is how governmental competition promoted innovation. Consider, for example, the corporation. Initially the fact that obtaining a corporation charter required a legislative act encouraged the holders of charters to bribe legislators in order to prevent new competition. As the corporation increased in popularity, the number of applications naturally increased. In response to these conditions, the industrial states gradually developed the legal form of a corporation from a limited charter under state supervision to a general incorporation by administrative act supervised by market forces. The move to general incorporation was directed toward creating equal opportunity and meeting the pragmatic need to lessen graft and limit legislative workloads to manageable proportions[19].

Since fees from incorporation could be an important component of state finance, states competed to entice corporations to relocate by providing the desired innovation in the corporate form. For example, in 1879, J. D. Rockefeller created the trust as a device to obtain a production cartel in oil refining and for some time this trust form of business organization was widely imitated. Eventually, however, it was successfully challenged as a monopoly under common law in the Ohio courts, and the parallel rise in demands for federal antitrust legislation created the need for an alternative form of business organization. The holding company, made legal by the state of New Jersey in 1889, became the legal alternative.

The development of the corporate form of enterprise would have been much slower if granting corporate charters had become a power of the the federal government. With federal control of charters the agrarians, who at best were suspicious of corporations, would have slowed the development. In fact, the institutional structure of capitalism exemplified by such movements as general incorporation and compulsory education developed much faster in an environment of state competition in the emerging national market than would have been the case with a centralized design. Support for government innovations promoting industrialization was initially concentrated in the states that specialized in manufacturing, but in time, as other states industrialised, they imitated the innovations of the original manufacturing states.

Second Hundred Years: 1890-1989

A brief survey of the evolution of the political economy in the second hundred years is in order to provide background material for the analysis of discovery, invention and innovation. During this period the economy was transformed from manufacturing to services and from national markets to international markets. In addition, as its scope expanded, government grew to become a large component of the political economy. Finally, the rate of discovery, invention and innovation accelerated during this period.

During the second hundred years services displaced manufacturing as the dominant economic activity[20]. The term services covers a variety of economic activities such as government, wholesale and retail trade, transportation and public utilities as well as consumer and producer services. The magnitude of the growth of these activities is reflected in the fact that today over 70% of the work force is employed in services.

When examined closely, however, this change from manufacturing to services is less than that implied by the figures. The definitions of services tend to overemphasize the shift from manufacturing since many services support the production, marketing and distribution of goods. Because centralized plants require goods to be distributed to the consumer, approximately 25% of the employed are in distributive and retail services. These distributive services consist of transportation, communication, utilities, and wholesale. The growth in employment in these services started in the first hundred years, as manufacturing began to expand. An even more recent growth area has been producer services which include finance, insurance, real estate, and corporate law. Also, the popular image of the growth of services as the growth of consumer services such as fast food franchises is totally incorrect. Actually, employment in consumer services has declined slightly because the decline in personal servants in private households has dominated this category.

Another major economic change that has taken place during the second hundred years is America's growing participation in the international economy. Improvements in transportation, communications, and the efficiency of capital markets have resulted in significant progress in integrating the world economy into a single market. Through applied science inventors created new forms of transportation–the airplane, auto and truck– and made improvements in all forms of transportation. These inventions and improvements have provided a much greater choice in speed, flexibility, and cost of transport for people and goods. In comparison with the first hundred years, current transportation is much faster and less costly.

Simultaneous with advances in transportation inventors and innovators have made numerous improvements in communications. Advances in telephone technology, for instance, have made the telephone a major instrument of communication. At first, given the telephone's initial high cost, businessmen were the primary users in local business communications. But with the falling costs of this device and service, individuals have made increasing use of telephones for personal communications. Falling costs, then, have greatly increased the volume of phone conversations by businesses and individuals alike, especially in the realm of long distance communication.

As a consequence of the advances in transportation and communications, an increasing number of goods have become items of world trade, and national economies have become more specialized. Improved communications have also enabled multinationals to manage activities worldwide, and low transportation costs mean that production for world markets can take place in almost any location. Following the example of the integration of world asset markets, entrepreneurs are beginning to use the advances in communications to integrate global markets for services. Given continuing advances in transportation and communication, economic incentives will promote a single, integrated, global political economy; however, progress is periodically interrupted by wars, trade restrictions in depressions and recessions, and numerous attempts to create lasting cartels.

The transformation of the American political economy from a system based originally on agriculture, then on manufacturing, and finally on services has completely transformed the social structure of this country since the 1790s. Currently most people live in metropolitan areas, where as a result of rising real incomes, they have experienced an increase in leisure. The creation of the automobile has made possible a much more mobile lifestyle, and as a consequence individuals now participate in a wide range of work, consumption, and leisure activities dispersed over a wide area. The telephone further supports this dispersed lifestyle by allowing individuals to participate in many activities without having to be physically present at a particular location. Finally, the advances first in radio and then in television have saturated society with new media.

Another important feature of the evolution of the political economy in the second hundred years to consider is the growth of government[21]. This growth of government as a regulator and provider of public goods was a major component in the previously discussed growth of services. In 1890, government constituted only 0.8% of all employment. By 1977, however, government employment, including jobs in public education, was over 25%. As a component of the gross national product, all levels of government have collectively grown to over 30% .

Because in the original design the concept of judicial review was supposed to keep the government limited to the original social contract, the Constitution, the expansion of the scope of government from limited to comprehensive might appear contradictory to the intent of the Constitution. And as modifying the Constitution was intentionally made difficult by the requirements of the amendment process, gaining authority for such growth by amendments would have been very difficult. But since the *Marbury vs. Madison* the Constitution has become the Constitution as interpreted by the Supreme Court. Hence the growth of government has been accommodated not by amendments, but by Supreme Court interpretations. For example, regulation of manufacturing was accommodated by the Supreme Court, after some initial reluctance, interpreting the commerce clause to cover manufacturing.

The growth in government occurred in three main periods-the Progressive Period from 1895 to 1914, the Great Depression of the Thirties, and the Post-War Period. The assumed tasks of government during these periods can be classified into two broad categories: regulation and the production of public goods. In regulatory tasks the government attempts to produce higher performance than would have been the case without regulation. Examples of government regulatory behavior include regulation of the environment and stabilization policy, which is the use of monetary and fiscal policy to reduce economic fluctuations and promote growth. Public goods, on the other hand, are goods and services, such as the military or education, for which the market solution is considered insufficient.

In the first period of government growth, the regulatory role of government was greatly expanded by the Progressives, a middle class reform group which organized to address the issues resulting from the transformation of the American economy from an agricultural economy based on small farms to a manufacturing economy led by giant corporations. The Progressives perceived actual market performance as considerably less satisfactory than the theoretical ideal of self-regulating markets. They considered child labor and the condition of women in the workplace undesirable and foresaw that the high accident rate in factories and mines could become politically destabilizing. For these reasons, the Progressives organized at the state level to regulate such issues as child labor, women in the workplace, and worker safety and health.

In addition, the Progressives believed the ruthless conduct of industrialists in creating the major corporations threatened the ideal of fair competition in the marketplace. Reformers addressed this issue of a competitive market at the federal level by the use of the Sherman Antitrust Act to prosecute business combinations, the passage of the Clayton Antitrust Act prohibiting predatory practices in business, and the establishment of the Federal Trade Commission to monitor competitive behavior.

Although Progressives pursued their objectives vigorously, the resulting legislation represented a compromise between the goals of the Progressives and the interests of business. The Pure Food and Drug Act, enacted to improve the sanitary conditions of the stockyards, for instance, was needed by the packers in order that they might to sell beef in Europe[22]. And it was major banking interests that pushed through the Federal Reserve System, thus establishing a central bank to act as a lender of last resort. This establishment of the Federal Reserve System in 1914 marked the federal government's first step in assuming the task of economic stabilization.

In the Depression the federal government expanded its role in stabilizing the economy[23], a role made law in the Economic Employment Act of 1946, and asserted its obligation to provide economic security. Under mainstream academic economic theory of the time, the federal government had no stabilization role in a depression because the economy was perceived as a self-regulating mechanism. Under the prospect of social unrest or, even worse, a move to radicalism, however, Roosevelt felt compelled to act in the traditional improvisatory strategy of government. The basic idea behind much of the Depression legislation was that by preventing competition, government could preserve the status quo firms. Since the growth of the large corporations, industry leaders had proposed the idea of ruinous competition, that is, the idea that industry could compete so hard that all members of an industry would be ruined. The proposed solution to industrial overcapacity during the Depression, then, was to cartelize each industry in an effort to restrict production, raise prices and restore profits.

The concept of cartelization was implemented in the Industrial Recovery Act which granted firms in each industry the right to form cartels. To obtain Labor's cooperation this act granted Labor the right to unionize. The move to total cartelization of the economy was stopped by the Supreme Court's decision that the Industrial Recovery Act was an unconstitutional delegation of legislative power. Despite this decision, Labor obtained legislation for its part of the IRA, and individual industries such as oil, trucking, and airlines succeeded in obtaining government regulation to lessen the force of competition. The banking industry similarly obtained specific legislation limiting competition between banks, and in agriculture the policy of parity prices and restrictions to planting acreage was initiated.

In addition to its effort at cartelization in the Depression, the federal government began providing public goods which previously were considered private, such as credit to industry, farmers, and home owners and, through the creation of the Tennessee Valley Authority, electricity. The government's regulatory tasks were expanded to include soil conservation, securities and exchanges, and banking.

Finally, in the Depression the government created programs to promote individual economic security. One example of such programs is the Civilian Conservation Core, set up to provide employment for young men; another is the beginnings of the joint effort between business and the state and federal government to provide unemployment insurance. In the same time period the passage of the Social Security Act of 1935 marked the initial recognition of a federal welfare role, which initially provided aid to the needy.

After World War II the government continued to expand its role as a regulator and provider of public goods[24]. Since the war the United States has assumed the role of protector of the free world in response to the Soviet threat and to this end, the accelerated rate of discovery and invention has continually boosted the cost of keeping the military technologically superior. However, now with the collapse of the communist empire and the end of the cold war, this component of government expenditures should start to decline.

Later in the Sixties, which like the Progressive Era was a period of relative prosperity, politicians enlarged government roles acquired in the Progressive Era and the Depression in hopes of obtaining a higher quality of life. These politicians cited market failures to provide public goods with positive social value and the control of externalities as a rationale for expanding government. The production of public goods believed not sufficiently supplied by the market, such as education and research, was expanded. In a similar growth of the government's welfare role, a concerted effort was made to reduce the number of people below the poverty line by supplying specific services such as food stamps, housing subsidies, medicine and aid to needy children. The needs of the aged, too, were met by an expansion of social security benefits and the creation of Medicare, a program that provides medical service to the aged.

A major new regulatory concern that has developed in the postwar period is the environment. While government regulation of the environment dates back to the establishment of national parks and forests in the first hundred years, the new emphasis is now on controlling environmentally harmful externalities of previous economically successful applications of the natural sciences. The Clean Air and Water Act exemplifies the resulting legislation aimed at regulating various forms of environmental pollution. To administer this and other environmental legislation, the Environmental Protection Agency was established in 1971.

The government's tasks as an umpire and as a stabilizer of the economy, then, have expanded greatly since the Progressive era. Initially, in the Progressive Era, the role of government was to act as an umpire to make business competition a fair game. In the 1960s, however, this role was expanded to include all individuals regardless of race, sex or creed, ensuring that they could obtain equal opportunity in employment, or be free from discrimination as consumers. A consequence of acknowledging in law the government's role as stabilizer of the economy in 1948 is that the status of the economy, primarily perceived in terms of employment before elections, has become an important campaign issue in subsequent presidential elections. Politicians in order to solve the plethora of tasks which they assumed have made governmental organization progressively more complicated. One type of choice creating this complexity has been the preference after 1887 for creating independent government agencies to regulate business rather than rely of private action under state common law. The resulting growth of business regulation through administrative agencies has created a patchwork organizational structure for regulation. Early regulatory agencies such as the ICC generally regulated a single industry. The Post World War II regulatory agencies followed a new pattern of generally regulating a single activity – product safety, worker safety, and the environment – throughout the economy.

The provision and production of public goods has also developed a very complex organizational structure. Originally, the federal government and state governments were thought to operate in separate spheres of influence, and this separation was defined by the 10th amendment. In the first hundred years, in such areas as education, this division blurred somewhat because all levels of government assumed educational tasks to perform. By the 1960s a bewildering array of decentralization schemes among the federal, state, and local governments emerged. One mechanism for the new decentralization has been the federal grant to finance state and local tasks like urban renewal. The resulting structure of government has produced a great increase in activities for which all levels of government perform some of the tasks.

More than one level of government can also perform tasks in regulatory activities of government. For example, to clean up the environment, the federal government mandated that state and city governments take action. Also, to promote safety and health in the workplace the federal Occupational Safety and Health Administration, OSHA, was established over state workman's compensation programs which in turn were superimposed over the earlier state common law.

The growth of new multilevel activities of government has been sanctioned by the Supreme Court's interpretations of the 10th amendment. For example, with the 1985 *Garcia* decision[25] the Supreme Court has now interpreted the 10th amendment to imply that decentralization of government states is to be determined by the national political process. This totally political criterion means that the complex multilevel organization is likely to remain the norm for any new governmental activities.

Acceleration of Discovery, Invention and Innovation

The third feature of the evolution of the political economy during the second hundred years to consider is the acceleration of the rates of discovery, invention and innovation. Since 1790, the rates of discovery, invention, and innovation has continually accelerated, especially during the second hundred years. As was previously pointed out the foundation of this acceleration was the creation of public education and the research university during the first hundred years.

In the second hundred years the rate of discovery has increased due to public funding of all types of research. In 1887 the Hatch Act initiated public funding for university research in agriculture. The impetus for expanding the scope of such public funding was the success of applied research in World War II in producing such technology as the atomic bomb. The National Science Foundation was created to publicly fund research in most fields through a peer group review system of grant proposals[26]. Public funding increased in response to the Soviet cold war threat and the specter of Soviet technological domination raised by Sputnik. Up to the present, high levels of expenditures for basic discovery have been sustained by a growing realization of its importance in military and economic competition.

Public funding of all types of research created the research university as the institutional arrangement to accelerate the rate of discovery. Political support for the creation of the research university with a teaching role was a coincidence of the need to educate the baby boomers and the need to dominate science. Major state-supported universities were accordingly transformed from teaching institutions to research institutions through the expansion of PhD programs. The PhD student became both a junior researcher and the ubiquitous TA or teaching assistant, an arrangement which enabled the faculty to spend more time on research and at the same time teach more students by using TAs.

The magnitude of this expansion in research can be indicated by the fact that in the second hundred years the number of universities conducting research has increased by an order of magnitude, and there are more scientists living today than have lived in all previous times. In addition to universities, government laboratories and some private laboratories are engaged in basic discovery. While the expansion has occurred in all fields of study, funding has been highest in fields corresponding to the perceived potential for advances in military weapons, economic competition and social value.

At the same time that the rate of discovery accelerated, institutional innovations directed toward obtaining a more systematic approach to invention and product improvement accelerated the rate of invention. The need to develop a better methodology for invention was the fact that starting with the electric light and the telephone in the latter part of the 19th century, a characteristic of most new inventions has been that they require considerable applied research to reap their full economic potential[27]. Also, as scientists accumulated knowledge in the natural sciences, formal university training to grasp this knowledge made inventors more efficient in improving existing inventions and creating new inventions[28].

Accordingly in the second hundred years to promote applied research as the basis of invention and product improvement, corporations developed specialized corporate research and development laboratories staffed by university-trained specialists. By World War I, over fifty corporations had laboratories[29], and today most corporations, intent on improving existing products and inventing new ones, conduct organized research and development at various levels, from basic knowledge research for major breakthroughs to the most mundane forms of applied research to create minor improvements in existing products.

International competition has led to a continual effort to accelerate the rate of invention. For example, as was discussed in Chapter One, basic research and marketplace invention have very different incentives and information flows. To speed up the development of market applications from basic research, public and private decision makers have taken many steps to bridge this gap. One example has been public funding of applied research by agencies such as the Department of Defense. This funding is much greater than funding for basic research. Unlike basic research, most of this research is directed at achieving specific objectives.

With new forms of technology resulting from applied research, new forms of property rights are required to promote continued inventive activity. A recent advance in this area has been the extension of copyrights to software and chip masks, which are designed to manufacture integrated circuits, and another has been a joint public and private effort to create the applied research consortium, a development made possible by a relaxation in the antitrust laws. The consortium, which is sometimes supported by public funding, enables firms to pool the risk involved in long-term development of research ideas. An example of such a consortium is MCC, which was formed to promote inventions in hardware and software.

The private sector has also made innovations to speed up the development of market applications. Major corporations, for example, now fund university research with various agreements as to the allocation of the resulting economic value. Also, financiers have innovated a new type of finance known as venture capitalism to promote new ideas in new firms known as start ups. Venture capitalism has developed in a modern environment of research universities surrounded by high tech firms. The new firms emerge either from university research or develop as splinter groups from established firms.

This institutional structure of new product development is in constant flux. Currently there is a recognition of the need to improve the transfer of knowledge from public government research laboratories into private economic products. Developing an efficient transfer mechanism from public laboratories to private firms is as difficult as developing such a mechanism between research universities and private firms.

The new institutional arrangements have supplemented rather than replaced the old. The lone inventor continues to remain active creating new useful products frequently without a deep knowledge of the underlying technology. Such inventors exploit the considerable body of practical discovery concerning the technical components generated by the expansion of the industry. An example of a product developed this way is the Apple personal computer created by a technician in the computer field. Also, corporate research remains active in improving existing products and developing new products.

The acceleration in invention has led to a continual stream of new products being introduced into the political economy. And along with increased invention, the rate of innovation has necessarily increased simply because of the vast quantity of new products and applications to consider. Providing a good index of the high rate of innovation are the numerous innovations based on the invention of the digital computer, since advances in computer hardware and software have had an impact on all aspects of basic knowledge accumulation, invention, and innovation.

In mathematics, for example, two researchers used 1100 hundred hours of computer time to prove the four-color map conjecture[30], which implies that only four colors are required to color a US map such that no two adjoining states have the same color. The proof is controversial because it deviates from the standard of simple, easy-to-verify arguments.

Similarly, in some branches of science advances in theory require computations that can only be performed by supercomputers. Meteorology, for example, now produces fluid dynamic models of the atmosphere which can consist of tens of thousands of equations. A practical result of such computer studies is much improved world weather forecasting. In invention itself, the major innovation the computer has created is the great reduction in trial-and-error testing through computer-model simulation. Chemical reactions can be simulated with a computer, greatly reducing the effort it would otherwise take to find new compounds with economic value.

Likewise, advances in computation are the basis for advances in the design and production of products. In designing products such as automobiles, wind tunnel tests and stress tests can be conducted on the computer design by using computer simulation programs. With this method numerous potential designs can be evaluated at much lower costs since no prototypes are required. In addition, the application of computers is also generating major innovations in automating production processes. For example, a recent step in the progress towards automation of production has been the creation of a computer language by GM called MAP which creates communication protocols that enable the machines in the factory to communicate with one another.

Equally significant are the ways in which computers have improved administrative paperwork. For example, before the computer, businesses such as insurance typically organized this paperwork as an assembly line, with each individual responsible for a minor component of the processing. The computer was initially used to expedite the functioning of these routine administration paperwork assembly lines, but with advances in computer terminals and software, all of the operations involved in processing insurance claims and customer service functions were organized in software such that a single customer representative could handle all of the customers' problems. Similarly, with advances in accounting software, a firm can analyze the profitability of its operations in increasing detail. This has led to reorganization and improvements in incentive systems to link rewards for managers with the profitability of their functions.

Using computers humans are also making major innovations in their decision-making processes. With the creation of decision-support systems and analytic software tools, the computer is promoting a shift from intuitive to analytic decision making. The availability of large volumes of transactions data provides corporations with new ways of making more sophisticated analyses of alternatives. Among the computer tools that are used to analyze these alternatives today are linear programming, expert systems, and spreadsheets, developed from the fields of operations research, artificial intelligence and finance, respectively. Before the creation of software-decision-support systems, middle managers performed the analysis themselves and passed reports up the chain of command. Now, however, in increasing numbers of businesses, all of these data analysis functions can be performed by a small number of staff people at the corporate headquarters, using increasingly sophisticated software. The resulting innovation in corporate structures has been a decrease in the number of levels in the corporate hierarchy.

The shift to a more analytic framework for decision making using information technology has not resulted in an instantaneous improvement in performance. In fact much of the expenditures in information technology has been wasted because management failed to carefully consider how the information system was supposed to improve performance. Improvements in the decision framework using technology require a careful redesign of the tasks to be performed using the new combination of humans and technology. Genuine innovations in the use of new inventions such as computers and improved methodology have advanced the rate of innovation. The new computer based analytic approach to decisions increased the rate of innovation by enabling decision makers to consider many more alternatives than were possible with the earlier intuitive evaluation methodology Currently, this new analytic framework for considering alternatives is primarily employed in business and government.

In addition to better evaluation technology, some industries have made improvements in the intuitive improvisatory strategy for innovation. In agriculture, for instance, the adoption of a better innovation implementation strategy, the separation strategy, has accelerated not only the process of discovery and invention but also the process of innovation itself. Starting with the Morrill Act of 1862, legislation created a system of agricultural experimental stations for research and a system of county agents to transmit their successes to farmers. Agricultural research stations now experiment with production techniques such as fertilizer application as will as inventions such as new seeds and equipment. This means that innovation itself is accelerated by the adoption of a separation strategy wherein innovation as well as invention benefits from good, statistically designed experiments[31].

Evaluation of Discovery, Invention and Innovation

With the growth of the international economy in the second hundred years, the ability of a nation's individuals, firms and government to discovery, invent and innovate has become has become one of the most important aspects of international competition. One consequence of the growth of government during this period is that the overall performance of the political economy has become as dependent on governmental performance as private-firm performance. Therefore, the ability of government to innovate has become as important as the ability of private firms to innovate. Also, individuals must innovate new lifestyles to integrate new technology into everyday life.

With the growth of a competitive world economy, the rates of discovery, invention and innovation have increased throughout the developed world. An important aspect of this change has been that the rate of invention has accelerated much faster than the rate of innovation.

The current disparity between the invention rate and the innovation rate can be attributed to two factors. The first factor is the empirical fact that the rate of discovery which promotes invention has accelerated relative to the rate of discovery which promotes innovation. That is scientists working in the natural and biological sciences have made much larger gains in understanding natural and biological processes than scientists working in the social sciences have progressed in their efforts to produce knowledge about social behavior.

The advances in scientific knowledge have led to the creation of simulation programs which enable inventors to evaluate the performance of alternatives without experiments. The advances in the social sciences, however, have provided fewer tools for analyzing alternatives, and the poor forecasting performance of most social science simulation programs limit their usefulness. This means that to evaluate an alternative for innovation, an empirical implementation is generally required. This is true even for many innovations in production because current knowledge in social sciences is insufficient to simulate accurately the impact of alternative organizations and incentive systems on worker and manager performance.

A second factor contributing to the disparity between the invention rate and the innovation rate is the empirical fact that the methodology of invention has advanced relative to the methodology of innovation. Currently most invention experimentation has advanced from trial-and-error to more systematic methodology. In contrast, innovation experimentation is still considered too expensive in most cases with the exception of experimentation for agricultural production innovations. The strategy for innovation implementation is primarily an intuitive, improvisatory strategy. This means that business, governmental, and lifestyle innovation remain largely the creations of the heroic figure of the entrepreneur.

The fact that invention has increased relative to innovation is clearly evident in the fields of medicine and law. In medicine, advances in physics and chemistry have been transmitted through biology to provide a much better understanding of how the cell works, and this advance has lead to a great increase in inventions for curing illness. This ability to cure illness, however, has advanced much faster than innovations in the delivery of affordable medical service to all citizens. Knowledge of the economics of medicine has advanced slowly, and systematic experimentation in medical delivery systems and incentives would be prohibitively expensive.

Similarly, in the field of law, fast-moving technology such as computation and communication creates new social problems faster than the judicial system can set precedents or legislators can create appropriate legislation. One such example is the resolution of the conflict between privacy and efficiency in the computational analysis of transactions data. Although the study of law is now being integrated with the social sciences, as in the law and economics movement, advances in the social sciences are insufficient to provide judges with tools for setting precedents more rapidly or to enable legislators to create legislation more quickly.

The fact that the innovation rate lags behind the invention rate has an important implication concerning the ability of the US to compete internationally. Before considering this implication explicitly it is first necessary to examine whether the US rates of discovery, invention and innovation are currently competitive worldwide.

First, consider the rate of discovery which promotes invention. With the creation of large numbers of research universities and a competitive funding mechanism, the US has developed an institutional arrangement and incentive system which produces an internationally competitive rate of discovery. One measure of this success is the number of Nobel Prizes awarded US scientists. Maintaining this internationally competitive rate of discovery requires maintaining the appropriate levels of funding for both research and research facilities. Organized lobbying groups are likely to succeed in this effort. Nevertheless, as the rest of the world advances towards the economic levels of the US, it is unrealistic to think that the US will dominate every field. In a specialized world economy the US need dominate only the fields promoting inventions in the areas of US comparative advantage.

Second, consider the rate of private invention and manufacturing innovation. These two must be linked together because of the empirical fact that the success of most inventions requires a consistent inventive effort at product improvements and a consistent innovative effort to develop an efficient production process. In this regard, the US has developed an internationally competitive rate of new inventions with the promotion of applied research and the development of corporate research and development facilities. As was the case in discovery, the advance of the specialized world economy means the US can not expect, nor should attempt, to dominate invention in all industries, only those US-specialized industries.

The aspect of private invention and innovation in which the US is currently seriously deficient is in product improvement and manufacturing innovation. One reason Japanese firms are better at product improvement than their US rivals is that Japanese managers have a longer planning horizon than their US counterparts. A second factor of the 1980s was the higher relative US interest rates. This has led to the reluctance of US managers to make the massive research and development expenditures necessary to transform an invention which is technologically feasible into an invention which is economically profitable.

Equally important, the US has lost its position as the leading innovator in manufacturing. Given the size of the US market, the US manufacturing philosophy at the end of World War II emphasized large scale production in very large batches. Changing the production process from one product to another was expensive and time consuming. Because land was inexpensive, the US manufacturing managers had little incentive to reduce the size of manufacturing complexes by reducing the large inventory of parts and work in progress.

After World War II US managers were smugly complacent concerning the superiority of their manufacturing processes and neglected to imitate the improvements of their foreign rivals. For example, when W. Edwards Deming, a statistical quality control innovator, advocated US manufacturers adopt his approach, he was soundly rebuffed. In contrast, his ideas were enthusiastically adopted in Japan. Generally, in established US industries manufacturing became a neglected activity staffed by the least capable managers, who emphasized cost reductions in the existing production facility and avoided major risky innovations. In addition, management-labor conflict resulted in stiflingly rigid work rules.

In contrast, after World War II the Japanese innovated a new production methodology based on small production batches and the flexibility to quickly switch production from one product to another. Because land was expensive in Japan, managers had powerful incentives to reduce the space required for manufacturing by greatly reducing inventory in parts and work in progress. This lead to the just-in-time manufacturing philosophy which holds as the ideal: zero inventory. In order to accomplish the goal of just-in-time inventory, Japanese manufacturers had to place tremendous emphasis on quality control which resulted in Japanese product superiority. Moreover, the Japanese developed an education system to produce more qualified workers than the US. Also, Japanese managers provided workers with an incentive system to achieve large advances in productivity through worker participation in decision-making.

The US did not respond to the Japanese challenge until the Japanese electronics firms had driven their US rivals from the consumer electronics business. Although the US currently leads in some areas of manufacturing innovation such as applications of computers to design and production, the US lags in many areas such as applications of robots and management-labor relations. Currently US government and corporate policy makers are well aware of the deficiencies of US manufacturing innovation and are beginning to make a concerted effort to correct these deficiencies by adapting Japanese innovations to US culture. This imitative effort is likely to take several decades and should result in the surviving US manufacturing industries being equal to their foreign rivals in innovation.

Outside of manufacturing innovation, the rate of innovation in the US is comparable to rates in other countries. The US is an innovation leader in the application of computers to all aspects of society and the individualistic ethic of Americans encourages rapid lifestyle innovations such as telecommuting.

To give the US a competitive edge in discovery, invention and innovation, a great deal more needs to be done than simply solve the problems causing the US to lag behind Japan in manufacturing innovation. Currently the rate of innovation is the bottleneck which prevent the US from more rapidly absorbing the high rate of invention. As long as the rate of innovation remains a bottleneck there is little point in trying to increase the rate of invention or the rate of discovery which promotes invention to still higher levels. Accelerating the rate of innovation would give the US a tremendous competitive edge, even though other nations would quickly imitate US successes in this endeavor[32].

An important step in promoting innovation is to increase the rate of discovery which promotes innovation. While the rate of discovery in the the social sciences and business disciplines has accelerated, the increase has not been as great as in the natural and biological sciences. And although the government funds considerable research, most of its resources go toward creating a public feedstock of ideas for the private development of inventions. As the benefits of social sciences research are general and frequently controversial, the social sciences lack powerful interest groups promoting their interests and consequently, current funding for the social sciences is only 5% of the NSF budget[33]. Nevertheless, tools to analyze alternatives are required to improve the methodology of innovation beyond the intuitive improvisatory strategy. And one way to bring about an improvement in developing tools to analyze alternatives is through adequate funding for research promoting innovation.

Among the obstacles impeding an accelerated rate of discovery in the social sciences and business disciplines are the conflicts in scientific information policy. To promote discovery in social behavior, the approach of empirical science requires systematic observations. Specifically, scientists need to make systematic observations of all political economic behavior in order to discriminate much more rapidly between competing theories. Unfortunately, private collection of data is generally impeded by conflicts with proprietary considerations and privacy. Firms generally desire to restrict the flow of such information to prevent competitors from understanding their competitive positions. Between private parties, of course, information policy depends on voluntary release of information as modified by required disclosures. For example, producers of food products must disclose the ingredients, including additives, on the label. Private parties typically filter voluntarily released information in such a way as to promote their own interests, and because of this such information does not generally constitute a representative sample.

In contrast to private individuals, the legislature has the power to obtain information to formulate legislation. Unfortunately, this power is primarily used for attention-getting investigations which are more useful for educating the public to the need for action than for providing data for the systematic study of the respective behavior. The administrative agencies since the 50s have had the same powers as the legislature in obtaining information to accomplish their specified missions[34]. But while the administration collects and releases some representative samples of information such as census data, most of the information is collected only for administrative purposes. Consequently, most administrative data has little empirical value because it does not contain the desired variables or have sufficient controls to be considered a representative sample. In fact, for much business data, the government only releases aggregate data which masks rather than reveals behavior.

Given the severe restrictions on obtaining representative samples, most empirical research on political economic behavior is frequently carried out using data samples collected for oblique reasons: hypothesis testing is restricted to those hypotheses for which data is available. Naturally, this problem inhibits the development of the business disciplines as well as the social sciences. For example, an empirical management scientist can rarely obtain representative samples of data on the relationship between incentives and performance of middle managers across companies in an industry.

Resolving the conflicts in scientific information policy is just one step toward establishing a better information policy to promote innovation. An acceleration in the rate of discovery will produce an increasing number of analytic tools with which to investigate alternatives, and with the growth of computers and software, these tools will increasingly be part of decision support systems for all types of decisions. The tools will aid choice behavior, promoting innovation and imitation as well as better decisions in general by allowing the decision maker to better evaluate his potential alternatives.

But tools to evaluate alternatives are useless without the requisite data. For this reason, operational information policy should enable decision-makers to evaluate both the positive and negative aspect of decisions so that they can be internalized into decision making. As parties in market decisions have incentives to voluntarily disclose positive aspects, generally only negative aspects need to be considered for required disclosure policy. Operational information policy will lead to better decisions and additionally, will require less government regulation. At the same time, individuals need some privacy and must still be protected from abuse of information. Likewise, firms must retain protection from loss of trade secrets. Information policy must therefore balance the needs of scientific advance, operational efficiency, and privacy.

Even if observations are improved nothing will be learned if the variables under study do not if fact vary. Individuals, firms, and the government generally pursue their respective goals with the best current knowledge. This means that for many variables of interest, goal-seeking behavior can result in very little variation. For example, rival firms in an industry can each use very similar production techniques, organizations and incentives. This is especially true for the federal government, where equal treatment before the law and an aversion to experimentation inhibit variation. Private firms have conducted experiments in incentives such as the famous Hawthorne experiments[35], which were conducted by General Electric to determine what factors influenced worker productivity. Since World War II, the government has conducted a limited number of social experiments in such issues as peak load pricing and guaranteed income[36]. Currently the amount of variation in public and private policy (especially systematic variation from experiments) is far to low to promote discovery which in turn promotes innovation.

More variation in government policies could be obtained by a better criterion for government decentralization. With government activity gravitating to the federal level in the second hundred years, the amount of learning that take place through government variation has decreased. The usual justification for decentralization of government is to allow lower levels of government to choose the locally preferred policy. Another reason to consider decentralization, however, is when lower levels of government all have the same goal but each has a different theory about how best to accomplish this goal, decentralization promotes variation. In such a case, decentralization enables different government units to base their policies on the theory which they themselves believe to be correct. Moreover, with different policies controlling different regions, empirical evidence can be gathered concerning the consequences of the various policies. For policies where learning is really important however, the variations should be experimentally controlled in order to rapidly reject bad policies.

Finally, the rate of innovation would increase if the strategy to implement an innovation were made more systematic. For many innovation tasks this can be accomplished by shifted from the intuitive improvisatory strategy to the previously discussed separation strategy. The classic example of a separation strategy for innovation is agricultural production innovation where applied research is performed at agricultural research stations and the results are transmitted to the farmers by agricultural agents. For this latter strategy to be effective there must be a large number of individuals, firms, or governments with similar tasks and incentives to fund joint experimentation. To some extent this condition has been met in the joint research consortia. Another alternative is the creation of publicly funded research for the learning necessary to improve the performance of a new alternative in a repeated task.

Analysis of government

Improving the performance of government as an innovator requires more than just improving the predictive capacity of the social sciences and adopting a better innovation-implementation strategy. As a result of the 20th century expansion of its scope and size, government has become so complex that even a dedicated scholar can not hope to grasp all its details. The question to consider is what has been the impact of this complexity on the ability of the government to innovate–that is, to produce changes which improve political economic performance with respect to the true but unknown common weal.

As the scope and complexity of government increases, an empirical fact needs to be emphasized. The more the scope of government increases the less performance of the political economy is simply a matter of private market performance. Overall performance of the political economy has increasingly become a function of the acquisition and application of knowledge by government. For government action to improve the common weal, government must both generate a good objective function for a task and use an appropriate theory to select a good alternative.

The government's task of stabilization illustrates this point. In defining the objective of stabilization policy, what emphasis should be placed on stable prices, reduced fluctuations in GNP, and growth? Should these objectives be achieved by the policies proposed by proponents of cartels of the Thirties, monetarists, Keynesians or the new supply siders? To achieve a higher performance than the unregulated market can produce, the government must be able to both agree on a good estimate of the common weal and understand economic behavior well enough to achieve good results.

Given the importance of good government performance in the modern political economy, one might ask how the innovators of government growth hoped to achieve such performance. One technique which they have used to promote performance has been specialization. Given their limited cognitive abilities, legislators have specialized the vast scope of governance into committees and subcommittees, each assigned to a limited topic. Government officials have also specialized the bureaucracy into departments and agencies, each of which administers a narrow aspect of the government's activities. Consequently, particular legislative committees and subcommittees develop long standing relationships with particular departments and agencies.

The limited resources of legislators, moreover, has meant that increasing aspects of the legislative process have been delegated to the executive. For example, to ease its task of specifying legislation, Congress leaves the details of implementation to the bureaucracy. And since the creation of the Budget Bureau in the 1920s, the budget has been submitted to Congress by the executive. In addition, since the 1930s most legislation has originated in the bureaucracy and has been submitted as the President's legislative program to Congress for modification and approval. Recent attempts of Congress to regain some delegated power from the executive through legislation such as the Budget Reform Act have been at best only partially successful[37].

The principal idea for achieving bureaucratic performance, dating from the Progressive period, has been the idea of staffing bureaucracy with competent professionals. Also, during the growth of government efforts to improve performance have been directed at making government more representative. Examples of democratic changes are the direct election of senators, voting rights for women and minorities, application of the principle of one-man-one-vote, and numerous attempts to reform campaign contributions. Finally, some effort has been directed at trying to improve the professional performance of legislators. One example such effort has been the increase in the professional staffs of legislators.

Because the original system of checks and balances was designed to promote performance by limiting governmental scope, it is useful to ask to what extent does the current system of checks and balances promote good governmental performance. Effective government under checks and balances requires that the authority of the executive be distinct from the authority of the legislature. However, the use of independent agencies for regulation and the delegation of legislative details to the bureaucracy has made this distinction ambiguous. For example, an independent regulatory agency is under limited control by the executive and operates as a quasi-legislative and quasi-judicial agency with the power to both define details of administration and make judicial types of decisions. In general, the legislature, having delegated details of legislation to the bureaucracy, believes it has a mandate for legislative oversight to ensure the bureaucracy is adhering to legislative intent. This conflicts with the separation of powers because the President is the chief executive of the administration. Such conflict becomes intense when the President and the majority of Congress are from different political parties.

The courts have sought to ameliorate the ambiguities in the separation of powers. Through precedents, the courts have sought to separate the quasi-legislative and quasi-judicial activities of agencies and to define bureaucratic due process. They also have placed some limits on the notion of legislative oversight.

Today if one party does not control both houses and the Presidency, conflict over policy exerts a strong political restraint on governmental action. The President will have trouble passing his legislative program, and the President will threaten to veto the legislative program of the majority of Congress. But in this political arena, analysis is secondary to political considerations. The more controversial a bill the more it is likely to be carefully analyzed to obtain supporters. For noncontroversial issues, legislators lack incentives for careful analysis of the underlying issues in creating legislation. And to simplify this analysis, legislators usually do not carefully consider tradeoffs unless they are required by law such as the Environmental Impact Statements.

In the conflict between the executive and the legislature, the executive is evolving a policy review of the bureaucracy by the executive branch. As the administration has grown, the executive oversight of administration action has shifted from the cabinet to the Office of Management and Budget, OMB. As this role of the OMB evolves, the OMB is proceeding to create a policy review of bureaucratic actions. With executive order 12291, the Reagan administration required analysis of the economic impact of new regulations[38]. While this was primarily instituted as a political device to stop the flow of new regulations, it marks an important step toward greater analysis before taking action.

As currently constituted, the judicial review of the courts does not promote professional analysis since it functions as a check to ensure that government action does not exceed the limits specified by the Constitution as interpreted by judicial precedent. From 1887 to 1937 the court interpreted the 14th amendment due process clause economically, denying that states had the right to interfere with contracts. In addition, when the ICC was established in 1887 the Court felt compelled to review all bureaucratic decisions brought before it. In a 1910 decision, the Supreme Court decided to review only bureaucratic matters of law and not matters of fact. This decision was later incorporated into the federal and state administrative procedures acts. Thus the courts, which do not provide a professional review mechanism for government themselves, leave that task to the legislature. The legislative review, however, is a political rather than a professional review, and professionalism remains a secondary issue in maintaining popular support. As they are currently designed, none of the three branches of government has both the incentives and the expertise to conduct a professional review of legislative and executive action.

The important issue is whether more powerful mechanisms such as a professional review are required to achieve good governmental innovation performance as defined by the criteria of Chapter One. To repeat, the first criterion is a competitive criterion that the change persist and be imitated by other governments. This criterion could be applied to government at all levels because even the federal government exists in a competitive international environment. The second criterion is that the change exhibit the properties of general benefits, consistency and efficiency. This second criterion is a stronger criterion because it can be applied immediately without waiting to see if other governments imitate the change. For this reason the second criterion will be considered in the analysis in this section.

Let us consider first whether the selection of alternatives by government has the property of general benefits. The starting point for this analysis is to consider the voter. To maintain himself in elected office, a politician must obtain the support of the majority of the voters in his district. Currently, the government has become so complex that a detailed understanding of the total range of government activities is beyond the reach of even a specialist. Government activities generally have a skewed distribution of cost and benefits; that is, either the costs or the benefits are concentrated in a narrow group in society. Because he possesses limited resources and cognitive limitations, the voter will generally focus his attention on those government actions that most directly impact him with sharply focused costs or benefits. In addition, a voter may have strong opinions on a small number of emotional issues such as gun control or abortion, and finally, a voter may have opinions on the small number of political issues to which the media is giving extensive coverage. Given his cognitive limitations, the individual voter has very little incentive to invest major effort to understand all the issues.

The estimate of the common weal by the individual voter, then, contains a bias for concrete, immediate results from government on a limited number of issues. The voter does have an incentive to try to influence government by voting or supporting a lobbying group to the extent that the perceived benefit exceeds the perceived cost. But the ability to organize voters into lobbying activity is inhibited by the fact that voters stand to receive benefits of the lobbying effort independent of whether they participate. Given this free rider problem, organized lobbying is more likely to pursue concentrated rather than diffuse interests[39].

To be reelected, the legislator[40] focuses his efforts on satisfying the concentrated interests of his constituents and the interest groups funding his campaigns. Furthermore, to promote his electability, a legislator seeks the committee assignments which are of the greatest importance to his constituents. The desire by voters for concrete, immediate results creates incentives for legislators interested in being reelected to focus on actions that will produce these concrete results before the next election. Because mistakes can be corrected in the next legislative session, the important reelection point is to achieve something for the next campaign.

The election incentives thus tend to produce committees staffed with members promoting concrete, immediate results for concentrated interests. This staffing of the specialized legislature has a profound effect on the estimate of the common weal. In a complex government, accurate estimates of the true tradeoffs between alternative government programs are not available. For example, the tradeoff between a particular military procurement and a particular social program is only vaguely estimable. To ease the problem of tradeoffs, legislators have adopted behavior such as logrolling[41] and the legislative code of get-along-by-going along which greatly reduce the need to make difficult tradeoff calculations. In the budget, the need to consider tradeoffs is eliminated in the practice of incremental budgeting, where each government activity receives a fixed increase. Also, if possible, the different groups compromise by giving each special interest the desired good. Only when a situation becomes a political crisis, such as the budget deficit has now become, do legislators have to make difficult decisions. In a complex government, then, the estimation of the common weal tends to be a collection of particular goods for particular groups. This bias will be called the *intensity bias.*

When government assumes a new task, frequently the controversy surrounding this action places the process in the public eye through the media. Legislators, therefore, have incentives to ensure that the generated political objectives and choice mechanism do not obviously cater to a particular interest group. However, most government activities eventually recede from the public eye to become primary concerns of the organized interest groups most directly affected by government action.

Because the task is generally organized through an improvisatory strategy, this means the political objectives and the choice mechanism are typically shaped over time by the interest groups with the most political influence.

The intensity bias impacts the bureaucracy as well as legislation. In any regulated industry such as insurance, the producers generally have more concentrated interests than the consumers, and thus the regulatory agency comes to promote the interests of the producer. However, exceptions exist. For example, because the consuming states have much more political clout than the producing states, federal regulation of natural gas benefits the consuming states. Similar problems exist in the provision of public goods. For example, congressmen in order to maintain employment at aircraft production facilities in their districts have been known to force the Pentagon to purchase unwanted military aircraft. Over time, then, the provision of public goods tends to benefit the organized interest groups most directly affected.

In additional, legislators make compromises which tend to result in complex legislation containing collections of particulars appealing to various interest groups. The tax code prior to 1986 exemplifies this tendency by the sheer volume of detailed special provisions. The complexity of the new tax code will, undoubtedly, increase over time. The more complex the government structure becomes, the more such government tasks take place outside the public eye and under the watchful eye of organized interest groups. In short, the more complex the government becomes, the greater is the intensity bias.

Consequently, all government actions taken together have the property of general benefits in the very limited sense that they represent the compromises of the legislators. But, the intent of defining the property of general benefits was to apply it to each legislative and bureaucratic act. Clearly, many government actions have very questionable general benefits. For example, efforts by the government to prevent barrier islands from undergoing change by nature are of questionable long term value[42]. These efforts provide temporary benefits to the small group of land owners. For a second example, agricultural policy may temporarily impede inefficient farmers from leaving agriculture at the expense of higher prices for the general public. Still another example, government promotion of forestry in Alaska has resulted in resources being sold to the Japanese below cost. Finally, trade restrictions generally promote the economic interests of inefficient declining industries at the expense of the consumer.

The complexity of government creates other equally important deficiencies in political choices. An example is a lack of consistency. For the purposes of this book, consistency will be defined as follows: if one committee makes choice A, implying C is in the common weal, and a second committee makes choice B, implying not C is in the common weal, then A and B are inconsistent. The most obvious example of this type of inconsistency appears in subsidies to the tobacco industry at the same time that health warnings must appear on cigarettes. Either the promotion of tobacco is in the common weal or it is not. It is similarly inconsistent for the government to exclude the oil industry from the superfund cleanup of toxic wastes. There are numerous other examples of inconsistency in government policies such as the inconsistencies in risk policies such as health risks. Such inconsistencies exist, of course, because government is specialized and responds to organized interests that are affected by each specialized government activity. Moreover, there are few incentives to eliminate inconsistencies.

Another deficiency in political choice is the lack of strong incentives for efficiency[43]. Governmental officials do take periodic steps to improve governmental efficiency. For example, bureaucrats imitate private sector innovations in data processing and office automation. The President periodically appoints presidential commissions to recommend steps to improve governmental efficiency, and over time many of their recommendations are implemented. Also, economic competition between political economic units states promotes efficiency. For example, to attract corporate relocations, state and local politicians need to provide essential services such as quality education and at the same time keep the tax burden down.

Nevertheless, voters do not have strong incentives to demand efficiency in the goods they receive from government. Indeed, the intensity bias of voters can even result in voters demanding governmental inefficiency. Consider the problem of too many military bases, a situation which is inefficient in providing the public good, national defense. Voters who might lose their jobs demand that their elected officials keep their base in operation. For many years such voter pressure prevented the Pentagon from closing unwanted bases. Finally, Congress passed legislation whereby inefficient bases would be closed by a blue ribbon panel thus exonerating elected officials from the dirty deed. In general, once government produces a service, however inefficient, it thereby creates a vested interest with a strong incentive to maintain the service.

Finally, the consequences of defining innovation as an improvement in governmental performance measured by a true, but unknown common weal must be considered. Legislation and policy are either directly or indirectly based on theories of behavior which generate an estimate of the common weal. If the theory underlying legislation is rejected, the political decision-making structure should respond by quickly revising the respective legislation or policy. Prompt action would have the properties of general benefits, consistency and efficiency with respect to a new widely accepted theory.

As David Hume pointed out in the 18th century[44], the truth of any theory cannot be validated by experiments without unwarranted assumptions. And as Popper notes[45], empirical science is one directional in pointing out the fallacies of theories rather than confirming theories. In a subject such as mechanics, basic revisions in theory may occur at infrequent intervals, but most government policies are based at least in part on practical wisdom theories which become discredited quickly.

Currently, the almost total reliance of political mechanisms on the system of checks and balances creates great difficulties in changing legislation once the theories upon which legislation is based are found to be invalid. To illustrate this point, let us consider an example of Depression legislation. As was pointed out, in the 1930s, banks, trucking, airlines, and the oil industry all succeeded in obtaining government regulation restricting competition. This legislation was based on a practical-wisdom understanding of the concept of ruinous competition. In the expanding global economy after World War II, however, the training of a greatly expanded group of professional economists increasingly discarded the concept of ruinous competition in favor of the more formal theories of competitive markets. Analysis performed by these economists demonstrated that restrictions on potentially competitive industries were not in the common weal; and in fact the restrictions on bank competition contributed to the post-World War II recessions. By the end of the Sixties both conservative and liberal economists were in rare agreement that the restrictions should be removed.

After a ten year effort the industries mentioned were deregulated in 1980, not so much because the legislators were convinced of the shift in economic theory but by the public concern for inflation[46]. This slow response of government reflects that fact that regulation creates vested interests who have incentives to resist change by influencing votes of the respective subcommittees. Given the complexity of government, only issues of major importance become campaign issues for the voters. Wide voter concerns for inflation forced Congress to deregulated to increase competition.

In this battle between professional opinion and vested interests, variation played a key role. Two airlines existing completely within particular states were unregulated and empirically demonstrated lower rates than nationally regulated carriers. In trucking, too, the transportation of agricultural commodities was unregulated and similarly demonstrated lower rates. But as the deregulation story indicates, because of the intensity bias government responds slowly to new knowledge that decreases the value of entrenched positions of vested interests.

The evaluation of the benefits of deregulation to society will take several decades. The cost of deregulating financial institutions would have been much less if deregulation had occurred earlier and careful consideration had been made on the interaction between deregulation and government guarantees of deposits. Also, no check was made on airline mergers or the number of gates an airline could control at an airport. A long time perspective is needed to evaluate these mistakes against the long term benefit of greater competition in financial institutions, transportation, and telecommunications.

Proposed Plan

The United States has numerous current problems which reduce its competitiveness worldwide. For example, primary and secondary schooling is deficient compared with our rivals. Also, the deficit needs to be reduced in order to reduce the cost of capital. The present system will struggle with these problems[47] over the next few decades. The focus of this book is the advancing the rate of innovation such that the political economy will achieve better performance in solving future problems.

During the past one hundred years, the rates of invention and discovery which promotes invention have accelerated relative to the rates of innovation and discovery which promotes innovation. Also, as the former rates are promoted by powerful vested interests, their rate of advance will probably remain competitive worldwide. To more effectively compete worldwide, the latter two rates must be increased. Any plan to promote innovation and discovery which promotes innovation must pay special attention to the defects in the process of political choice.

To accelerate the rate of innovation four steps will be proposed. First, more research expenditures are needed to promote discovery which leads to innovation and to support a shift to a separation strategy for innovation. Second, a better information policy must be established to resolve the conflicts between empirical science, operational decision making and privacy. Third, a new decentralization criterion must be established for government to increase the amount of learning thus promoting more innovation in government actions. And fourth, governmental reforms are required to promote the selection of alternatives which have the properties of general benefits, consistency and efficiency.

Of the four steps the last is by far the most difficult to achieve. Improving governmental performance by providing voters with better information has limited potential. Even if an inexpensive and comprehensive information system could be provided for every voter, the voter would generally find the cost in terms of the time it would take to fully understand the issues to be much greater than the rewards of such knowledge. An alternative to such a solution, however, is to reconsider the system of checks and balances which the founding fathers created to reduce the defects in the political process. In this book the system of checks and balances will be modified by imposing a professional review mechanism on governmental actions.

As the four steps to promote innovation will require major changes in the present arrangements, it might appear to the reader not to be worth the cost. The justification is simple: Any nation which fails to match its competitors in efficient social organization will be at a disadvantage in competing economically and militarily.

Given the level of intensity bias in the current design, many may wish to improve only private innovation for their personal gain. This approach, however, would neglect the most important potential improvement in performance, for the impact of government growth has been to link the performance of the public and private sectors. Overall performance of the political economy, then, is very much a function of government performance. For example, the severity of the Depression was magnified by the failure of the Federal Reserve System to prevent the collapse of the banking industry[48]. Since that time the increased number of tasks assumed by government and the accelerated rate of discovery and invention have greatly increased the complexity of most decision making. Initially, most decisions were made by isolated farmers under conditions of slow technological change. But as the market economy developed during the first one hundred years, decisions came to be primarily two party contracts regulated by common law.

In the current political economy, however, all two-party contracts have potential third-party effects which become the potential concern of government legislation and administration. In the second one hundred years, decision making has become increasingly collective, and at the same time the rate of technological change has accelerated. Because the problems of managing government increase more rapidly than the number of tasks that must be coordinated, the overwhelming number of tasks which the government has assumed has made government decision making much more complex. Furthermore, this increase has increased the complexity of decision making for all participants in the political economy. Private decision makers must now consider more government-imposed conditions and incentives and must also be able to cope with a much higher rate of change in which all actions are subject to possible government regulation. The impact on the firm has been that many more resources are required for decision making. Improving public innovation performance, then, will by extension improve private innovation performance.

Notes and References 2

1. For a discussion of the economic conditions at the end of the colonial period see any text in American economic history. For example, Hughes, J., 1990, *American Economic History, 3rd edition*, (Scott, Foresman and Company: Glenview)

2. The weaknesses of the confederation are detailed in the *Federalist Papers*

3. Taylor, T.,1955, *Grand Inquest*, (Simon and Schuster: New York)

4. Roche, John P., 1961, The Founding Fathers, *American Political Science Review*, Vol 55

5. For example see modern copy of: Hobbes,Thomas, 1640, *Leviathan*, Locke, John, 1705, *Of Civil Government* and Rousseau, J., 1763, *The Social Contract*

6. While the concept of a judicial review is not specifically spelled out in the Constitution, a majority of the framers believed the judiciary had this power. See Sundquist, James L., 1986, *Constitutional Reform and Effective Government*, (The Brookings Institution, Washington)

7. Historical Statistics, Ser. F10-21, p139. For a summary see North, D.C., 1974, *Growth & Welfare in the American Past*, (Prentice-Hall,Inc.: Englewood Cliffs, N.J.)

8. North, Douglass C., 1965, The Role of Transportation in the Economic Development of North America, *Les grandes voies maritimes dans le monde XVe-XIXe siecles*, Paris

9. Veysey, Laurence R., 1965, *The Emergence of the American University*, (The University of Chicago Press, Chicago)

10. Gunderson, G., 1976, *A New Economic History of America*, (McGraw-Hill: New York)

11. Rosenberg, Nathan, 1972, *Technology and American Economic Growth*, (Harper Torchbooks: New York)

12. See reference 11 and for comment on McCormick see: Bruchey, Stuart, 1975, *Growth of the Modern American Economy*, (Dodd, Mead and Company: New York)

13. For a survey of the innovations in business organization, see: Chandler, Alfred D. Jr., 1977, *The Visible Hand: The Managerial Revolution in American Business*, (The Belknap Press of Harvard University Press: Cambridge)

14. That is not to say that the social rate of return on government's efforts to promote transportation faster than the market would have, was necessarily positive. For example, in the case of canals, see: Ramson, Roger, 1964, Canals and Development: A Discussion of the Issues, *American Economic Review*, LIV, NO. 2 15. Gwynne-Thomas, E. H., 1981, *A Concise History of education to 1900 A.D.*, (University Press of America, Inc.: Washington)

16. Posner, R. A., 1986,*The Economic Analysis of Law, 3rd Ed*, (Little Brown, Boston)

17. Hazard, L., 1971,*Law and Changing Environment: History and Process of Law*, (Holden-Day: San Francisco)

18. See reference 10

19. See: Berle, A. A. and G. C. Means 1933, *The Modern Corporation and Private Property*, (The Macmillan Company, New York), Hurst, J. W., 1970, *The legitimacy of the Business Corporation in the United States, 1780-1970*, (University press of Virginia: Charlottesville) and Liebhafsky, H. H., 1971, American Government and Business, (John Wiley & Sons, Hew York)

20. Stanback, T., P. Bearse, T. Voyelle and R. Karasek, 1981, *Services the New Economy*, (Allanheld, Osmun Publishers: Totowa, NJ)

21. For a theory concerning the growth of government, see: Higgs, R., 1987, *Crisis and Leviathan: Critical episodes in the growth of American government*, (Oxford University Press: New York)

22. For a contrast between the early liberal view of the Progressive period with the radical view compare: Hofstadter, R., 1955, *The Age of Reform*, (Alfred A. Knoop: New York) with Kolko, G., 1963, *The Triumph of Conservatism*, (The Free Press: New York). Subsequent writers have tended to balance the two positions.

23. For a survey of the depression see: Mitchell, B., 1960, *Depression Decade: From New Era through New Deal 1929-1941* (Rinehart & Company, Inc.: New York)

24. For example for a survey of the Johnson's Great Society see: Divine, R. A. (ed), 1981, *Exploring the Johnson years*, (The University of Texas Press: Austin) and for a survey of the increase in regulation see: Weidenbaum, M.L., 1977, *Business, Government, and the Public*, (Prentice-Hall, Inc.: Englewood Cliffs)

25. Garcia vs San Antonio Metropolitan Transit Authority, 1985, 105 S.Ct

1005. This decision reflects the cumulative changes in Supreme Court interpretation of the 10th amendment.

26. Stine, Jeffrey K., 1986, *A History of Science Policy in the United States, 1940- 1985*, Report prepared for the task force on science policy, Committee on Science and Technology, House of Representatives, (U.S. Government Printing office, Washington, DC) NSF report on science policy

27. See reference 11

28. For example, by 1880 a majority of engineers were university trained. See: Ray, John, 1979, The application of science to industry, in Oleson, A. and J. Voss (eds.) *The Organization of Knowledge in Modern America, 1860-1920*, (The John Hopkins University Press: Baltimore)

29. For a reference to the development of industrial research see: Lewis, W. D., 1967, Industrial Research and Development in Kranzberg, M. and C. W. Pursell Jr. (eds), *Technology in Western Civilization Vol II* (Oxford University Press: London). For an interesting account of the development of industrial research at GE and Bell see: Reich, Leonard S., 1985, *The Making of American Industrial Research: Science and Business at GE and Bell, 1876-1962* (Cambridge University Press: Cambridge)

30. Appel, K. and W. Haken, 1977, The Solution of the Four color-map Problem, *Scientific American*, October, pp 108-121

31. A famous article of the social rate of return of agricultural research is: Griliches, Z., 1958, Research Costs and Social Returns: Hybrid Corn and Related Innovations, *Journal Political Economy*, October, pp 419-31.

32. Dertouzos, Lester, and Solow point out that one of the deficiencies of American manufacturing managers is the failure to innovate new strategies. See: Dertouzos, M., R Lester and R. Solow, 1989, *Made in America: Regaining the Productive Edge*, (The MIT Press: Cambridge)

33. Knezo, Genevieve, 1986, *Research Policies for the Social and Behavioral Sciences*, Report prepared for the task force on science policy, Committee on Science and Technology, House of Representatives, (U.S. Government Printing office, Washington, DC) NSF report on science policy

34. The ruling in *US v. Morton Salt Co.*, 338 U.S. 632 gave administrative agencies this power.

35. For a discussion, see, for example: Roethlistberger, F. J., 1941, *Management and Morale*, (Harvard University Press: Cambridge)

36. Hausman, J. A. and D. A. Wise (eds), 1985, *Social Experimentation*, (The University of Chicago Press: Chicago)

37. Sundquist, James L., 1982, *The Decline and Resurgence of Congress*, (The Brookings Institution: Washington)

38. Grubb, W. N., D. Whittington, and M Humphries, 1984,The ambiguities of benefit-cost analysis: An evaluation of regulatory impact analyses under executive order 12291, in Smith, V. K. *Environmental Policy under Reagan's Executive Order: The Role of Benefit-Cost Analysis*, (The University of North Carolina Press: Chapel Hill)

39. Olson, Mancur, 1971, *The Logic of Collective Action*, (Harvard University Press, Cambridge)

40. One of the first public choice theorists to consider individual incentives in politics was Downs. See: Downs, A. 1957, *An Economic Theory of Democracy*, (Little, Brown, Boston) The analysis presented is a variant of bounded rational choice theory. For a readable, nontechnical survey of current social choice theory the interested reader might consider: McLean, Iain, 1987, *Public Choice: an Introduction*, (Basil Blackwell: Oxford)

41. Logrolling does not necessarily converge to a stable solution. For example, see: Dummett, M., 1984, *Voting Procedures*, (Clarendon Press: Oxford)

42. Dolan, R. and H. Lins, 1987, Beaches and Barrier Islands, *Scientific American*, Jul, pp68-77

43. The usual connotation of government efficiency is waste in the sense of using an excessive amount of resources to produce a given amount of output. A second type of inefficiency is producing an inappropriate amount of the public good. Niskanen argues that budget maximisation by bureaucrats leads to excessive amounts of public goods. See: Niskanen, W. A., 1971, *Bureaucracy and Representative Government*, (Aldine: Chicago)

44. See Hume, David, 1739, *Treatise of Human Nature*

45. Popper, Karl R., 1959, *The Logic of Scientific Discovery*, (Basic Book, Inc.: New York)

46. Derthick, Martha and Paul J. Quirk, 1985, *The Politics of Deregulation*, (The Brookings Institute: Washington)

47. For example, reference 31 contains a proposed program of innovations to improve manufacturing productivity.

48. Friedman, M and A. Schwartz, *A History of Monetary Policy since 1867-1960*, (Princeton University Press, Princeton)

Chapter 3

Microbinics

Introduction

The new design of government will propose major changes in the current government. For such major changes to gain public support they must, of course, undergo public scrutiny, and even if considered desirable, such significant governmental changes are likely to require a long gestation period before enactment in modified form as a compromise between the various factions. Consider, as just one example, the Freedom of Information Act of 1967 granting the individual the right to access government files[1]. After Watergate this legislation was strengthened in 1974 and 1976, but the demand for freedom of information, in fact, dates back at least to the election of 1912, when Woodrow Wilson recommended an "open files" policy towards government and corporate information[2]. Because the changes proposed in this book are major, the time frame for possible implementation in some modified form is estimated to be between 50 and 100 years.

A good design must take into consideration the technology which is likely to come into existence while the design is being implemented. The consequences of new technology constitute considerably more than just an increase in material output or the introduction of new goods and services. In the past, technological advances have created social problems, for example pollution, which have been addressed by an increase in political action. Likewise, future technology will undoubtedly create new political problems such as the decline of privacy and the impact of gradually automated economic activities on the income distribution. Technological change frequently results in social change by making feasible new alternative social arrangements. The growth of the factory and the manufacturing city in the 19th century is just one example. In the future, if the expected major reduction in communication costs relative to transportation costs occurs, this change in relative prices will create incentives for creating new types of organization that substitute communication for travel. Thus, the design for a new political economy must incorporate the possibilities of new forms of social organization and, additionally, resolve the forecasted problems of new technology.

With the current high rate of expenditures on research and development, most types of technology are advancing rapidly. The discussion of new technology which follows focuses on those types of new technology which will materially affect the design of the new political economy. The first criterion for considering a new technology is its potential impact on society. One measure of the first criterion is the size of the industry which develops to provide the new technology. A second measure of the first criterion is the number and scope of the innovations fostered by the new technology. The second criterion is the sooner the new technology is likely to have a major impact on society the more emphasis it should receive in the new design.

The most important technologies for consideration in the new design are electronics and communications[3]. Within these broad categories, the two technologies which commands the most interest are the rapidly developing technology of integrated circuits and the use of laser light as a medium of communication. In this book, these technologies will be called *microbinics*. The "micro" portion of the term refers to the fact that in both industries the component functions are being miniaturized. The "binics" portion refers to the growing use of the binary number system in both electronics and communication. While the communication system is just now beginning the transformation from an analog to a digital system, the transformation will be complete long before any possible implementation of the design proposed here can occur.

Microbinics meets the two criterion for materially affecting the design. Microbinics is a large, growing industry. Advances in integrated circuits that led to corresponding advances in the power of digital computers have already fostered numerous, major innovations in mathematics, science, engineering, administration, and decision making. As will be discussed, further advances in microbinics will foster many additional innovations in automation and social organization. Microbinics meets the second criterion as the time frame for major innovations is from 1960 until perhaps the middle of the 21st century.

Many other areas of technology are also making rapid advances. For example, genetic engineering[4] promises new biological methods of manipulating plant and animal characteristics beyond the established method of selective breeding. Progress in this area could greatly increase agricultural production to feed the world's population. Additionally, genetic engineering is creating biological mechanisms for manufacturing products and processes such as the manufacture of human inferon with genetically altered bacteria. Comparable advances in physics and chemistry will continue to spawn a continuing sequence of new technologies. All of these new technologies will foster innovations in the production of the technology. Some will promote innovations in the application of the technology.

Nevertheless, what makes microbinics the technology warranting detailed consideration is the vast scope of human activities for which advances in microbinics promote major innovations. For the reader to grasp the impact of microbinics on society, a survey of the current developments in microbinics is presented in this chapter.

Binary Representation

For everyday use, humans much prefer the decimal number system to the binary number system, since patterns in strings of zeros through nines are much easier to recognize than equivalent patterns in the much longer strings of zeros and ones. For technological implementation, however, the binary number system is superior to any other number system. Technologically its great advantage is that each digit or bit can be physically represented by only two states, and there are many physical ways of realizing the two states: by the opening or closing of a switch, by the north or south pole of a magnet, by the asymmetric response of a transistor to voltage differentials, or by the presence or absence of a pulse of photons. Because these two states can be represented by a 0 and a 1, microelectronic devices can be conceptualized as devices to manipulate sequences of 0's and 1's or binary numbers.

This raises the question whether binary numbers are useful in human activities. First, decimal numbers can be converted into binary numbers and vice versa. Thus, the individual can input decimal numbers into an electronic device, and the device can convert the decimal numbers into binary numbers for processing and then convert the results back to decimal form for perusal by the individual. It is much easier to convert from decimal to binary and then back again than to design a machine to process decimal numbers directly. In wordprocessing a similar transformation takes place. As each letter is imputed into the wordprocessor it is converted into a binary number. The conversion follows a standardized convention such as the ASCII convention. In the electronic device the binary numbers representing the letters are manipulated, and when output is desired the binary numbers representing the letters are converted back into letters. All languages can be represented by binary numbers[5] and thus can be processed by a wordprocessor. Similarly, algebraic equations and symbols can be represented as binary numbers and thus be amenable to manipulation by software[6].

Sound and images can also be represented as sequences of binary numbers. In digital sound and television the original continuous analog signal is converted to a discrete digital signal for processing and transmission. The last step, at the receiver, is to transform the signal back into analog. The quality of sound or television depends on the number of times per second a measurement is made and the word size, that is the number of bits, used to quantize each measurement. For telephone quality sound, the waveform of the human voice can be measured 8000 times a second and each measurement can be quantized by a byte, an eight digit binary number[7]. To achieve the high quality sound of the new digital stereo, the sound of a symphony orchestra or rock group can be measured from between 36,000 to 44,000 times a second and each measurement can be quantized by a sixteen bit number[8]. The digitization of a television signal requires a much higher sampling rate than voice. For example, the CCIR standard of 1981 specifies a sampling rate of 13.5 million times a second for the luminance signal and 6.75 million times a second for the two chrominance signals[9]. Each measurement is quantized by a byte. Lower quality television can be achieved by less frequent measurements and fewer bits for quantization.

The quality of a digitized static image, for example a digitized photograph or a computer graphics screen at a particular instant in time, depends on two basic quantities: the number of picture elements, which are called pixels, and the number of bits used to quantize the information at each pixel. The pixels are arranged in a rectangular grid such as 640 columns and 480 lines. The more pixels a graphic artist has to draw a digital picture, the more detail the artist can incorporate into the picture. The number of colors or shades of gray, which an artist can draw at each pixel, equals two raised to the power equal the number of bits used for quantization. Thus, if only one bit is used at each pixel, each pixel displays two colors such as black and white and if a byte is used, each pixel can display two hundred and fifty six colors.

Another factor which affects the quality of a raster cathode-ray-tube image is the number of times per second the image is redrawn. To avoid a flicker, the image must be redrawn from between 20 and 100 times a second[10]. The higher the intensity of the light the more frequently the image must be redrawn. The total amount of bits per second which must be manipulated to produce raster cathode-ray-tube graphics is the product of number of pixels, the number of bits for quantization and the number of times the image is redrawn per second. With advances in integrated circuits, the trend is to use more pixels in a finer grid, more bits to quantize more colors, and redraw the image sixty times a second for clearer, more vivid images without noticeable flicker[11]. Currently the technology of sound and video is in the process of transformation to the new digital technology. Once this transformation has taken place, there will be a common technological basis for processing numbers, equations, words, voice, or pictures. Thus, binary numbers can be made to represent important aspects of human thought and communication processes.

Integrated Circuits

From the perspective of mathematical logic, the states 0 and 1 can be equivalently represented by true and false. What this means is that Boolean algebra can be employed to construct mathematical operations on the sequences of binary numbers. For example, using the three basic Boolean logic operations of "and","not" and "or," logical circuits can be designed to perform all arithmetic operations[12]. Physical circuits can then be designed to realize the logical circuit using a particular set of electronic components. The fact that the logical design exists in Boolean algebra means that the logical circuit and its realization are separate entities. If a new better type of electronic component is created, for example a Josephson junction or optical switches using laser light instead of electricity, then under ideal conditions the abstract mathematical circuit can be realized in the new technology without modifying the logic.

Currently integrated circuit technology-that is, technology involving microelectronic circuits constructed on an approximately 1/4 inch square silicon chip[13]-is advancing rapidly. The present developments in this type of technology originate from the development of the transistor, a small solid state asymmetric amplifier, in 1948. This asymmetric amplification is what gives the transistor the two state capacity. The transistor was a major advance because it was both more reliable and required much less power than the vacuum tube which it replaced in most applications. At first transistors were individually wired into electronic circuits on modular circuit boards. The need to develop smaller and more sophisticated electronics for guided missiles created a demand for miniaturization of these modular circuits. In the 1950s inventors made a series of advances which resulted in a manufacturing process to construct an electronic circuit which could be manufactured on the surface of an approximately 1/4" square of silicon, a chip.

Economic competition has driven integrated-circuit firms to constantly strive to increase the number of components on a chip. As the electronic components are packed more closely the speed of the circuits embedded in the chip increases, because the electricity has less distance to traverse between components and transistors switch more quickly. Thus, increasing the number of components on a chip results in faster, more powerful integrated circuits.

Increasing the number of components on a chip also tends to decrease the production cost. The cost of developing a integrated circuit is dependent primarily on the number of steps in the process; hence, for a given technology, if the number of components doubles, the cost of producing the integrated circuit increases by a factor less than two. Both economic and physical factors, however, limit the number of components in an integrated circuit. The economic limit to the number of components in an integrated circuit is the number of components which can be produced profitably. Increasing the number of components in an integrated circuit decreases the yield rate, which is the fraction of defect-free integrated circuits in a production run. The economic limit is determined from the interaction of increasing economies of scale and decreasing yield rate. Economic competition drives integrated circuit firms to make continual advances in quality control to increase the yield rates.

The experience gained in increasing the production of a particular type of integrated circuit is another factor which results in a reduction production costs. An empirical rule in manufacturing maintains that as the cumulative output of a particular device doubles, the associated manufacturing costs fall by 20 to 30%. Economies of scale, onsite learning to increase yield rates, and the fact that new facilities use more advanced technology than older facilities all contribute to the fall in costs. If we consider the cost of producing a particular device, such as a 4M DRAM (dynamic random access memory) then we would expect this empirical law to hold. If we consider the cost from the perspective of electronic component the fall in cost is associated with both the decreased costs of packing the electronic components more densely and the manufacturing efficiencies of building particular devices. For example, from 1970 to 1977 the cost per bit of random access memory declined an average of 35%[14] per year and has continued to decline about 30% per year since 1977.[15]

The design and production costs of integrated circuits have created two broad categories of integrated circuits. Memory integrated circuits, for example DRAMs are known as commodity chips in that the primary goal is low-cost-mass production[16]. At the other extreme are integrated circuits for specialized applications. Here the primary objective is reducing the time and the cost to design the specialized integrated circuit. As more and more components are crowded into a single integrated circuit, the design problems in using manual labor to create integrated circuits have increased. Software programs called silicon compilers have been developed to design integrated circuits automatically to achieve a specified objective[17]. Consequently, with the advance in silicon compilers there is trend towards an increasing number of custom-designed integrated circuits.

Economic competition in integrated circuit design and production led to an annual doubling of the number of circuit elements contained in an integrated circuit each year from 1959 to 1972. This empirical relationship is known as Moore's law[18]. From 1972 to today, doubling the number of components has taken about 18 months, due to the greater effort required to package more components in an integrated circuit. By the year 2000 it may be possible to pack a billion electronic components onto a chip [19] . Increasing the number of components on a chip has led to a tremendous increase in demand. With lower costs the demand for goods using integrated circuits increases and this, in turn, creates a increasing demand for more integrated circuits. From 1960 to 1977 the annual number of transistors manufactured in integrated circuits has doubled 11 times.

As the number of components on an integrated circuit increases firms will encounter physical limits. One physical limitation is the wavelength of the electromagnetic source used in lithography. To obtain circuits more densely packed than is possible using light lithography, an electron beam can be employed, but at much greater capital costs. As the limits of current metal-oxide-silicon field effect transistors are reached, a new even more miniaturized quantum effects technology may be created.

At some point in miniaturization engineers will encounter fundamental limits based on physical science[20].

Besides integrated circuits based on silicon, engineers are working on other promising technologies. One example is replacing silicon with geranium arsenide, which facilitates faster operation. Another example is superconducting circuits such as the Josephson junction which are potentially 1000 times as fast as a silicon transistor[21]. Other faster options are optical devices using laser light instead of electricity and optical switches instead of transistors[22]. More exotic and possibly of much greater economic impact is the current research to build quantum effect devices at the molecular level.

Computation

Central to the design of most electronic devices is the processor, which can be contained in one or more integrated circuits. One consequence of the continual expansion of the number of components in an integrated circuit is that with each generation of electronic devices, more and more functions can be crowded into fewer and fewer integrated circuits. One example is the microprocessor, in which the processor is designed in a single integrated circuit[23].

The processor of an electronic device is capable of executing a sequence of instructions called a program. The program instruction execute circuits in the processor, for example, to add two binary numbers together. In designing and using processors there is a tradeoff between the hardware, that is the actual physical circuits, and software, that is the sequence of instructions. Consider multiplication of 3 times 2. To perform this operation one could use hardware by creating a multiplication circuit, or software by issuing a sequence of instructions to add 3 plus 3 in the addition circuit . Multiplication, in other words, can be performed in a hardware circuit, or by using software which employs repeated additions in an addition circuit. An example of the tradeoff between hardware and software is the current effort to develop reduced-instruction-set microprocessors, which achieve high performance with a small number of very fast hardware instructions[24].

Three important types of electronic devices employing processors are computers, communication equipment, and control devices. Consider first computers[25]. Most current computers are based on the von Neumann design in which instructions are executed in sequence and there is one channel between memory and the central processor unit, which may be supported by several subordinate coprocessors for specific functions such as numeric, graphic or signal processing. A von Neumann computer consists of a bus, that is a channel, connecting the central processor unit to fast access memory on integrated circuits, such as DRAMs and ROMs, and controllers for external devices[26]. Among these external devices are slower access memory devices such as magnetic disks, monitors, printers, and external communication ports such as modems.

Five categories of computers in terms of increasing speed and power are embedded, personal, mini, mainframe, and super. The cost and performance of each category is driven by advances in integrated circuit technology. For three decades the cost of computing has been falling some 20 to 30 percent per year. For each category of computer, the power in terms of the number of computations per second increases by a factor of ten every eight to twelve years.

When a computer is designed, it is, of course, based on existing integrated circuits, but by the time a computer is manufactured and enters the market the integrated circuits may be several years old. By the time enough software has been developed to make a machine a contender in the market, the integrated circuits may be over five years old. This means that the power of a new computer on the drawing boards may be eight times as powerful and have much more memory than the computer it will replace. The trend is for each generation of computers at a particular level to have the power of the previous generation of computers at a higher level. For example, the new personal computers based on the Motorola 68040 or the Intel 80486 processors have the power of the previous generations of minicomputers.

Along with advances in von Neumann type computers, new types of computers are being developed. One advance in computing is distributed processing-that is a network of computers of various sizes and capacities. A second advance is a parallel processor, which is a single computer with more than one central processor[27]. Inventors are making many experiments with alternative numbers of central processor units of varying computational power. The major software problem in such experiments is how to coordinate the processors to obtain maximum output. A new parallel computer concept which most closely resembles the operation of a human brain is a neural network[28]. This parallel concept, which can be implemented as hardware or software, shows great promise in pattern recognition, a capability in which the conventional digital computer does not excel.

Regardless of the potential power of a computer, its usefulness depends on the availability of software to perform the user's tasks[29]. An important aspect of making computers useful is reducing the cost of developing software. Consequently, one trend in developing software has been to create languages which reduce the programmer's effort in producing programs for the final user. The first machines had to be programmed in machine language, and the first advance from this stage was assembly language which substituted mnemonics for the binary machine instructions. In the 50s, two important languages, FORTRAN for scientific programming, and COBOL for business programming, were developed. FORTRAN enabled engineers and scientists to program equations in an intuitively natural manner. A compiler was constructed to convert the FORTRAN instructions into machine code for execution by the machine. Similarly, COBOL instructions provided a language of business operations which a compiler later converted into machine code.

Since the 50s a large number of languages have been created for a variety of purposes[30]. Many have incorporated computer science concepts such as structured programming to facilitate better program construction. The trend in languages is to facilitate more operations in fewer statements. One example here is the recent development of object-oriented programing languages. Nevertheless, the major impediment to improved software is that while the construction of computers has increasingly become automated, the development of software has so far remained labor intensive. Currently, considerable research effort is being made to develop artificial intelligence aids to programming to greatly increase the productivity of programmers.

Another important aspect of the usefulness of a computer and its applications software is the effort required for the user to learn how to operate the computer and to run the applications software. As the speed and memory capacity of computers have increased, a greater part of the increase has been used to make the computer and its applications software easier for the final user to use. The economic incentive for this trend is that the more productive a particular brand of computer and its associated applications software make the final user, the greater the demand for such a computer and software. The biggest advance in using computers is at the personal computer level. With the Xerox Star and later with the popular Macintosh, the user controls the computer and applications software by pointing at visual icons with a pointer called a mouse. Moreover, all applications software has the same format for operation. Thus operating this user computer interface becomes so intuitive that most users can quickly learn to use their computers and software without learning a long list of operating commands and with only an occasional reference to an instructional manual. But the mouse interface requires considerable memory and processor resources to operate. The value of the user's time, accordingly, drives the hardware and software developers to use cheaper memory and more powerful processors to make the operating system and software ever more "user friendly."[31]

Communication

Unlike computing, where major advances did not occur until the invention of integrated circuit technology, analog communication has made major advances without microelectronics since the invention of the telegraph and telephone in the 19th century. As the telephone has by now largely displaced the telegraph, let us consider only advances in the telephone[32]. Initially telephone operators manually connected calls. An early advance was the invention of automatic switches or exchanges. As the telephone system grew, the demand to transmit a larger and larger volume of calls between switches grew. To accommodate this increasing volume of traffic, communication engineers learned how to combine or multiplex phone calls for transmission. To transmit a signal of an increasing number of multiplexed telephone calls communication engineers harnessed higher and higher frequencies in the electromagnetic spectrum for use as communication channels. At the terminal switch, calls were separated to be sent to individual telephones.

Several factors currently promote the shift of communications from analog to digital. First with the rise in data communication for business an increasing amount of communication is inherently digital. Second, digital communication or pulse code modulation creates a signal for which, by using greater bandwidth, it is easy to maintain low error rates. The final factor is the fall in the cost of integrated circuits. With digital communication, integrated circuits can be used in various aspects of communication, such as multiplexing and switching. The cost of digital communication equipment falls along with the declining cost of integrated circuit technology.

With the shift in communication from analog to digital computation and communication fuse into a single technology for manipulating and transmitting information objects represented as strings of binary bits. As numbers, letters, symbols, voice and image can be represented by binary numbers this unified technology encompasses a broad range of human expression. A major benefit of this technological merger is that the cost of a single unified technology is much less than the cost of developing a separate technology for each type of computing and communication. Within this unified technology general purpose hardware can be controlled by special purpose software for alternative tasks.

Currently most local communication is analog while an increasing portion of long distance communication is increasingly digital. The first phase in the shift to a completely digital system is the proposed adoption of the communication protocol ISDN, integrated services digital network, which communication engineers developed to use the existing twisted wire or copper wiring between switches and home and offices[33]. For the basic ISDN service, the twisted wires will be converted to digital with two 64K D channels and one 16K B data packet channel. Higher capacity ISDN lines will also be available.

Beta tests for the ISDN network are being conducted currently and telephone companies plan to transform the telephone system to ISDN in the next couple of decades. While ISDN is designed to reduce investment costs to make the transformation from analog to digital, nevertheless the implementation of ISDN will require a major investment in ISDN communication equipment. From the perspective of the local telephone user the transformation to ISDN will be transparent. Until a local user has a need for ISDN digital services, he will maintain his analog telephone.

The extent that the basic ISDN line can integrate the various types of communication depends on making effective use of capacity of the two D and the one B channels. Shannon developed a theory of communication where the measure of the amount of information in a message is its randomness or entropy[34]. Thus to transmit a message an algorithm can be developed to compress the redundancy out of the raw digitized message at the transmitter and another algorithm can be developed to reconstruct the message at the receiver.

A second factor which can be employed to make effective use of communication channels is to transmit only the meaning of the message. The Shannon measure of information is a statistical measure which says nothing about the meaning of the message. The amount of information to transmit the meaning of a message is frequently much less than the Shannon measure of the total statistical information. For example, to transmit the meaning of a voice conversation none of the Shannon information in frequencies higher than 3400 cycles per second needs to be transmitted. Effective communication is achieved by compressing that portion of the raw quantized message which contains the meaning. For example, acceptable quality voice conversation can be achieve by compressing the 64000 bits a second of raw quantized telephone-grade conversations to 4000 bits a second.

Another active area of research is the compression of image communication. Depending on the convention, the raw quantized television signal varies from 90 to 216 million bits per second. Numerous algorithms have been constructed to compress raw quantized television signals[35]. The quality of the resulting transmission depends on amount of compression. Very poor quality dynamic image communication can be achieved in a transmission of only 56 thousand bits a second. This level of compression is achieved by selective transmission. For example, in communicating teleconferences the signal can be compressed by transmitting the background once and then transmitting the changes, that is the movements of the participants. Better dynamic visual communication can be communicated in a few hundred thousand bits per second. Through research, the quality of dynamic images at all levels of compression is improving over time.

The basic ISDN service can integrate voice, data, text, symbols, slides and low grade video communication in the existing local twisted-wire copper communication network. The ability to expand the capacity of this local copper network is limited because copper attenuates high frequency signals and installing the required number of amplifiers to boost the signals would be expensive. Fortunately, engineers have created a much higher capacity alternative. In the 1970s communication engineers, in their efforts to harness higher and higher frequencies of the electromagnetic spectrum to create channels with greater capacity, learned how to harness light as a media of communication. And at the same time scientists and engineers learned how to create glass fibers to serve as the media of communication for optical signals. The advantage of optical fiber over copper is that fiber can transmit very large capacity over local networks without amplification.

If fiber optics were used for local communication complete integration of all conceivable types of communication would be achieved. Currently optical fibers are being used for large volume communication such as long distance or between switches. Research in the capacity of fiber optics is doubling the volume of this medium each year[36]. In 1987, a single-frequency optical fiber had the capacity of 1.7 billion bits/sec. Much higher capacity is obtainable for shorter distances using multiple frequencies of light in a single fiber. With potential capacity in the 10s of trillions of bits per second, fiber optics has the capacity for communicating millions of voice conversations and thousands of television signals.

Application: Automation of production

An important application of microbinics is the automation of the production of goods and services[37]. From the perspective of economic production theory, automation is simply the substitution of capital for labor in the production of goods and services. This simplicity, however, masks an important difference between automation based on microbinics and earlier automation. Up to the advance of microbinics, most automation was mechanization which substituted machines for human labor. The role of man in the mechanized production process was to act as a flexible, intelligent controller for powerful, inflexible, stupid machines. The advance of microbinics, however, promotes the development of software to substitute for the human control function. As the proposed political-economic design is based on technological advances in the next 50 to 100 years, the design must be based on the impact of future automation. And the first step in creating a forecast of the impact of automation is to survey current automation advances.

The starting point for such a survey is manufacturing. Manufactured goods can be classified as discrete, as for example an automobile, and continuous–for example gasoline. Currently the production of many continuous goods is already automated. Consider, for example, chemical and petrochemical products. With an understanding of the chemical reactions required to produce the good, the chemical engineer can program the production process in a computer model. With sensors, the process can be monitored to detect deviations from the desired conditions. These deviations are fed into a computer which takes corrective action to maintain the process within the desired tolerances. In the production of many continuous goods, then, man's role has been reduced to watching the dials on the control monitor and taking steps in extreme conditions.

The automation of discrete goods, on the other hand, is currently proceeding at a steady pace. To discuss the progress being made and the problems to be overcome, let us divide the production process into four steps - design, parts manufacture, coordination, and assembly [38]. Progress towards automation is currently proceeding in a piecemeal fashion in each of these various components of the manufacturing process primarily in the aircraft, appliance, automotive, electronics, and heavy industries.

Increasingly, new products are designed using computer graphics packages, in a process known as computer assisted design, CAD[39]. The first CAD programs, which were developed in the mid-60s for automobile design, required a mainframe computer and cost in the millions. With the increasing power of computers at all levels, however, CAD programs have been created for smaller and smaller computers. Currently CAD programs are available on personal computers and work stations have sufficient power for serious design work. Because of greater capacity of smaller computer, the cost of CAD systems has fallen to the tens of thousand dollar range.

Simultaneously, software developers have made numerous advances in CAD software. One direction of advance has been to create specialized CAD software for each type of design such as mechanical, electrical circuits, integrated circuits and architecture. Another direction has been to greatly increase the capability of each type of CAD software. For example, in mechanical design the first CAD software represented solid objects only as outlines called wire frames, but with increased computing power, two and, later, three-dimensional solid objects could be represented, first as surfaces and later as solid objects.

CAD software has greatly increased the efficiency of designing products. In designing a component for a mechanical product, the designer can examine how the component fits with other components on his computer screen. Much of the increased efficiency of CAD lies in the fact that the designs reside in computer memory. Modifying a design can be performed easily by editing the original design. In design processes where styling is important, the ability to make modifications easily enables the firm to style without prototypes because the designer can view the three-dimensional representations of all the alternatives from any arbitrary orientation. In addition, firms which make many related products can greatly reduce their design costs by editing the design for one part in order to design a part for a related product.

CAD software has even automated steps in the design process. For most design problems the need for draftsmen is eliminated in that the engineer can make a rough sketch and the program provides the finished drawing. The current development of object oriented CAD software will provide the engineer with artificial intelligence tools to automatically analyze various aspects of a design such as consistency[40]. For some design problems, such as designing circuits and integrated circuits, CAD software can automate the design process itself.[41]. CAD can increase the efficiency of the design process by a factor of two to as much as thirty in the case of integrated circuits.

Before the CAD approach to design a coordination problem arose in designing products which underwent frequent revisions. As engineering drawings were shipped to various groups in a large organization, it could happen that the various groups were not all using the latest revision. With CAD, however, the design is in computer memory, to which all groups have direct access. Thus all groups will use the latest design automatically.

Much engineering analysis is necessary in order to implement a successful design. For engineering analysis which can be programmed, the analysis can be performed on the design as it exists in memory, thus eliminating the need for building a prototype. This process is known as computer assisted engineering, CAE[42]. One use of CAE is to compute standard physical characteristics such as the center of gravity. A second widely used CAE application is finite element analysis, which is used to compute the stress at every point in the structure. Another CAE use, primarily in aerospace and automobile design, is computing the aerodynamic properties of a surface such as a wing without the researcher having to resort to a wind tunnel test. The limit of CAE applications is the limit of the engineer's ability to accurately simulate the actual conditions the product must undergo with a software model. Within such limitations, however, CAE greatly aids design by enabling designers to eliminate obviously bad designs from consideration before building a prototype. The cost of considering a wide range of alternatives is thus greatly reduced.

The next stage of manufacturing is the production of the component parts. The economics of producing the component parts for discrete durable goods depends on the size of the batch or the production run[43]. The larger the run, the less the cost of setting up the run affects the cost of a single part. For runs of 200,000 or more, that is, mass production, it is efficient to have special-purpose machines for each operation. Each of these machines follows a single, defined sequence of operations. When the run is small say on the order of 10 to a few hundred, that is, batch production, it is generally not economical to use special-purpose machines for each operation. Consequently, general purpose machines, for example lathes or drill presses, are used for a variety of purposes in the production run.

Prior to 1960 a skilled machinist would set up a general purpose machine and perform the required operations, using his accumulated knowledge to create the part indicated in the blueprints. In thus using general purpose machines, the set up costs-for example, of orienting a part to a precise alignment-can be considerable. As these cost must be distributed over the number of parts produced, the cost of producing parts in a batch of, say, 10 could be 10 times the cost of mass production. To produce an individual part by exclusively using general purpose machines might be as high as 100 times the cost of mass production.

Currently the technology exists for automating the production of machined parts for batch production runs. This technology has been progressing since 1960, when electromechanical control devices were developed which would cut a part on a lathe automatically. The input to the controller, which determined the selected sequence of lathe cuts, was a paper tape. With the advance in microbinics, the ability to control general purpose machines advanced correspondingly such that by the 70s parts could be manufactured by an automatic machine shop where a computer controlled several general purpose machines. Such a automatic machine shop is called a manufacturing cell or a flexible manufacturing system, FMS[44].

In the 70s Ingersoll-Rand installed a manufacturing cell where a computer controlled six general purpose machines. In this setup the role of the operator is reduced to loading the raw material and removing the finished part, which has been manufactured by the sequential operation of several general purpose machines controlled by the FMS computer. In the 80s the Japanese company, Fanuc Ltd, constructed a plant which has thirty manufacturing cells producing parts for robots and machine tools[45]. Robots and unmanned delivery carts handle all material such that at night, a single operator can operate the plant using a closed circuit TV to monitor the operations. Currently inventors are working on software to link groups of FMS cells together for more complex operations.

The economics of the FMS parts manufacture are considerable. The FMS computer is much faster than a skilled machinist at setting up a part for processing. With a skilled machinist, over 90% of the time can be set-up time, but with control devices, the set-up time can be reduced to less than half. Thus, automated production makes much more effective use of the machines. Additionally, automated production requires few skilled laborers to operate; The skilled labor force can, in fact, be reduced by as much as a factor of ten. Hence, automated part production is more efficient both because it requires much less labor, and also because it makes more efficient use of the capital.

Another great advantage of FMS is flexibility. Since the cell is software controlled, all that is required to shift production to a new part is to change the software program. A manufacturer can quickly switch production in response to changing demand and can maintain much less inventory given the speed and low cost of switching production. This greatly reduces production costs to the extent that the cost of batch producing parts is falling to the level of mass producing parts. Finally, as Frost Co. discovered[46], another advantage of FMS is the reduction in rejects. In the Frost experience the number of rejected parts fell from one in four to one in twenty.

Efficiency in manufacturing requires careful coordination of the production and purchase of parts with their assembly into products, since many manufacturing plants produce a large number of products involving a much larger number of parts. Given variations in demand, optimally scheduling the machines to produce the required parts at the correct time is a very difficult problem. This problem is further complicated by the fact that the flow of parts and work in progress must be routed through the factory to reach the correct assembly station at the right time. Before computers came to be used for controlling the flow of parts, the solution was to have bins of parts and work in progress to ease the coordination requirements.

Maintaining large amounts of inventory to ease the coordination problem in manufacturing is expensive. First, the value of inventory maintained is an investment, which reduces the firms profits. Second, inventory takes up space and must be maintained. Finally, maintaining large amounts of inventory and work in progress inhibits quality control. If a part is found to be defective, the incentive is to replace the part with another rather than try to correct the problem creating the defect.

Concurrent with the advance in automation has been the tremendous reduction in inventory and work in progress in manufacturing. In the US, material resource planning programs have been introduced since 1968. Such programs use a computer model of the manufacturing process to schedule the machines and route the flow of production. The amount of work in progress inventory can thus be significantly reduced. The leader in inventory reduction is Toyota which introduced a Kanban system, which is a just-in-time approach to inventory control[47]. The goal in this inventory control system is to produce or order a part only if there is a specific demand to fulfill a production order. The ideal in just-in-time inventory control is to reduce inventory to zero.

The implementation of a just-in-time inventory control system is not possible without a great deal of attention to quality control. A defective part or work in progress means the final product can not be completed without stopping the assemblyline. To achieve zero inventory all problems in quality control must be solved. To achieve this end, Japanese management has organized workers into quality circles to solicit improvements from the workers and has implemented statistical quality control to detect defects quickly[48]. US and European managers are imitated Japanese advances in inventory and quality control.

In the factory, not only is the inventory of parts being greatly reduced but also the handling of parts is gradually being automated.[49]. In some older plants, parts can be handled as many as fifteen times before assembly; therefore, automation of parts handling helps both reduce inventory, labor costs, and errors. Automation of material handling is being accomplished with the installation of computer controlled delivery systems using such devices as conveyors, overhead rails and automatic guided vehicles.

An example of an almost completely automated parts handling operation is Apple's factory to assemble the Macintosh[50]. In this factory, the concept of just-in-time delivery is extended with a fully automated material delivery system. As parts arrive at the factory, they are unloaded into the appropriate containers. The delivery to the work stations is then controlled by a computer. Parts, with the exception of screws, are sent automatically as needed on a conveyor, overhead rail, or automatic guided vehicle. The screws are delivered manually a few times a month. The advantage of this system is that parts are handled only twice: when they are unloaded and when they are assembled. While the Macintosh has only 300 parts this technology will gradually be adapted to larger operations.

Progress in automation of assembly depends on advances in robotics. Unlike the humanoid robot of science fiction movies, the industrial robot is generally a mechanical arm with a gripper for a hand. The economic usefulness of a robot for assembly depends on the capabilities of the robot. In assembling a product a human worker uses his senses-mostly sight and touch-and motor skills to select the correct part and tools to assemble his component into the product. In assembling parts, the human is very efficient in aligning the parts and making small corrections for errors in alignment. If a minor contingency arises, the human can take corrective action.

To incorporate in a robot such capabilities as sensory perception, orientation in space, and the simple intelligence required to recognize the correct part and cope with contingencies is a major challenge to machine intelligence, a branch of artificial intelligence. For example, a human recognizes objects without much conscious effort. But a robot, in order to similarly recognize an object, must first have vision, perhaps a small TV camera. Then, a program must be constructed to recognize shapes. This seemingly innocuous task is now being solved with much research[51]. While current robots lack many human capabilities, they surpass humans in speed, strength, and predictability. Moreover, robots don't tire, take drugs or consume alcohol.

Economic application of robots in assembly depends on redesigning the product and assembly process to exploit the existing capability of robots[52]. One approach to reducing the need for sensors and machine intelligence in robots is to strategically position the materials such that the robot can perform a task without sensors or machine intelligence. One example of this approach is the application of robots in the 70s to weld and to paint automobile bodies. Because the frames are accurately positioned, robots can weld and paint without sensors or machine intelligence. Robots are efficient at welding because they are stronger than humans and robots are efficient at painting because they do not need clean air. Another example of reducing the need for sensors and machine intelligence is to replace bins of randomly oriented parts with dispensers which feed the parts to the robot with a precise orientation. With such a dispenser a robot can insert integrated circuits into a circuit board[53]. In such applications a robot is superior to humans because, once correctly programmed, the robot does not make mistakes.

As the machine intelligence of robots has advanced, their applications in automating assembly have increased. Consider, for instances, machine vision where advances are making applications economically in labor intensive applications which are repetitive such as identifying parts, inspecting quality, and precise measuring[54]. One example in the 80s of the use of machine vision to significantly advance the automation of assembly is Allen-Bradley Inc's plant to assemble as many as 999 types of contactors and relays which serve as elctromechanical starters and controllers for industrial electric motors[55]. To simplify the machine identification problem of which product to assemble, Allen-Bradley uses bar codes, which are similar to the bar codes on products in a supermarket. A laser measures the surfaces of magnets in the controllers to keep them within tolerances. The use of 3500 machine inspections has greatly improved quality control.

Progress towards complete automation of manufacturing is also taking place in integrating CAD, FMS, parts coordination, and assembly. The integration of these steps is called computer integrated manufacturing, CIM[56]. Progress in the various steps and in integration of the steps is promoted by the creation of universally accepted standards for model representation, software interfaces, and communication protocols. Standards are proposed by governments, engineering society committees, and corporations. If standards become accepted by the major players in a market, they usually become universal. Accepted standards expand the market by making equipment compatible and by enabling small producers to specialize in niche applications, knowing that their equipment or software will be compatible with other equipment and software. Such standards also facilitate integration of the various steps by creating standard interfaces.

One example of a standard which is likely to gain wide acceptance is the IGES graphics standard developed by the National Bureau of Standards, NBS, for CAD designed objects[57]. This standard facilitates the transfer of designs among design and manufacturing groups. Another new standard developed by General Motors is MAP, a protocol for machine communication in the automatic factory[58]. GM is currently installing MAP in its factories in order to link the entire production process electronically. MAP promotes the development of a paperless factory as all information flows whether for production or office functions can be defined in MAP. As many other firms are adopting MAP, it will soon be the standard.

An important aspect in advancing the various steps in automation and in CIM is making advances in automation software. Prior to the creation of microbinics the automation of human tasks was incorporated into the physical equipment itself. For example, in mass production, prior to software, special purpose machines could be connected to produce products with very little labor. This type of automation is known as hard automation. The great advantage of the new soft automation in which human tasks are programmed in software is flexibility. To produce a new product with hard automation, the plant has to be physically modified at great expense, both in terms of money and time. The goal of soft automation is that the output of the plant can be changed quickly by changing the software control program.

Programming human tasks is far from simple. Consider the task of a skilled machinist operating several machines to produce a part. To develop a high-level-programming language to program this task for an equivalent FM cell would, at the very least, entail developing commands to control the tools of the various machines in time and space. But this is only a small part of the skill of a competent machinist. A skilled machinist knows efficient sequences of operations to produce parts to the prescribed tolerances. For example reaming a hole follows drilling the hole. Also, sequencing of operations must consider such factors as heat buildup.

To integrate CAD with FMS requires the development of software which generates efficient sequences of operations to produce the part. Given the enormous range of parts and materials the knowledge of efficient sequences is immense and the creation of general purpose software is unlikely in the near future. Current software successfully integrating CAD with FMS achieves more limited objectives. John Deere, for example, has software to integrate CAD with FMS in the production of various types of sprockets[59]. By restricting the application to a narrow range of parts, knowledge of efficient sequences can be incorporated into the software. Another approach taken by NC microproducts integrates FMS with CAD as a simulation package which the designer uses to evaluate and revise tool paths[60]. In this manner the designer can input knowledge of efficient sequences manually.

More general integration of CAD with FMS requires much more software development. The IGES standard does not provide sufficient information such as tolerances to construct the part. To transfer sufficient information NBS and 200 companies are working to establish a Product Data Exchange Standard (PDES). To facilitate the integration of CAD and FMS design software needs to be defined in terms of recognizable objects rather than just lines, arcs, and points. In mechanical CAD such objects would include holes, bevels, and grooves. CAD software based on recognizable objects makes editing easier and easier the problem of generating the code to create the part. Knowledge of efficient sequences of tool operations needs to be developed as expert systems.

The progress towards complete automation requires the development of a giant database linking all aspects of design, manufacturing and office operations such as marketing and accounting. The database would contain the data to link all aspects of design and manufacturing. For example, included in this database would be the specification and tolerances for each production component. Control programs to control the physical equipment at each stage must be created. Creating these control programs will require new discoveries such as deep knowledge to construct dynamic models in real time and surface knowledge in the form of expert programs to implement rules in the manufacturing process. In addition, the database would also link manufacturing production with office operations, such as sales to better organize production runs. Thus a very important part of automation advance is discovery which becomes incorporated into software.

In addition, progress towards complete automation involves much more that simply automating the various human tasks in the current approach to design and production. Intense international competition among manufacturing firms is promoting both automation and nonautomation innovations in manufacturing. To obtain the full potential of both types of innovations requires a continual reorganization of both design and production.

Consider first the subgoal of quality. To implement just-in-time inventory control a very high level of quality control is required. However, a consequence of achieving the level of quality needed to implement just-in-time inventory control is more reliable products which customers prefer. Thus, the pursuit of quality becomes an important subgoal in itself not just to reduce the inventory costs but also to increase revenues through greater sales. This pursuit of quality has engendered innovations which have been enhanced by technological developments. For example, the innovation of statistical process control, which harnesses statistical reasoning to achieve quality, is enhanced by the creation of CIM databases. Also, developments in CAD and CAE promote the shift from the pursuit of quality by the use of inspectors at each stage of production to the pursuit of quality by designing quality into the product prior to production.

Another important example is the new emphasis on flexibility in production. Flexibility increases profits by enabling the manufacturer to eliminate inventory in organizing production to match demand. Thus the focus is on developing software controlled automation to produce a variety of products at a moments notice instead of hard automation to make a long production run for inventory.

Also, as the firm which first introduces a new product on the market has a competitive advantage over its rivals in gaining market share, firms have made numerous innovations to reduce the product cycle. One organizational innovation promoted by the Japanese and being imitated by the Americans is the creation of a design team for the design and manufacturing of each new product. Such design teams have members from all the firms's departments and have the authority to make decisions. This approach to new product development leads to much better coordination between design and manufacturing implementation and between product development and customer desires. In this regard, advances in CIM provide a framework for much better coordination in new product development.

The current advances in automation in manufacturing are concentrated primarily in the aircraft, appliance, automobile, electronics, and heavy machinery industries. Perhaps the most advanced is electronics. Integrated circuits can be designed automatically and the manufacturing of integrated circuits is being automated to increase the yield rate. Integrated circuits are installed into circuit boards automatically and IBM has automated the assembly of convertible computers in Austin[61]. Also, with advances in automation, diffusion to other manufacturing industries is also taking place. Many industries now use CAD to design products. For example, Nike uses CAD to design footware. Robots are now being applied to manufacturing in industries as diverse as candy makers, pharmaceutical houses, underwear manufacturers and plastics molders.

Entrepreneurs are also applying microbinics to automate production tasks outside manufacturing. In agriculture robots are being created to pick crops such as oranges. In biotechnology automatic gene synthesizers are being created. Automation is also being applied to the manipulation of physical objects in services. In printing typesetting has been automated. Utilities such as electricity, water, and sewage are continuous products whose production is similar to continuous-process manufacturing. Consequently, these types of activities are currently automated. In transportation, too, the operation of the propulsion system of ships has been automated. And in overnight delivery of packages, the sorting is performed by automatic sorters. Even automatic transportation systems have been created in controlled environments, such as the automatic vehicle system between concourses at the DFW airport. In wholesale and retail trade the handling of goods is being automated with the innovation of automatic warehousing and in medicine the performance of laboratory tests are being automated.

Application: Paperwork

Inventors and innovators are also applying microbinics to enhance the performance of individuals and institutions in tasks involving the manipulation and communication of information objects defined in terms of text, data, symbols, sound or image. Such tasks can loosely be called paperwork and given this broad definition, CAD and CAE are examples of paperwork. Advances in applying microbinics to enhance performance of paperwork tasks usually proceed from the development of standalone hardware and software to manipulate information objects to the development of communication protocols and networks to communicate information objects. What makes the application of microbinic standalone hardware to paperwork tasks an innovation is not the potential capacity of the hardware in itself but rather the invention of applications software to perform useful tasks and the reorganization of work to take advantage of the new software capabilities.

One of the first information-manipulation activities enhanced by microbinics has been so called number crunching or the manipulation of numerical data in standalone computers. Since the development of the mainframe computer, software has been invented for to perform administrative tasks such as bookkeeping. Transferring administration to mainframe computers has facilitated the programing of numerous routine tasks such as report generation. In addition, an expanding library of software has been invented to provide scientists and engineers with tools for analyzing scientific and engineering problems. Over time software inventors have expanded the scope of software to encompass most numeric processing tasks. For example, linear programming and statistical packages have been developed to aid decision-makers.

As a computer is vastly superior to a human at arithmetic, the creation of software encourages innovations to make numerous computer calculations beyond the capabilities of humans. Thus with accounting software, firms can compute the profitability of thousands of products-a Herculean task for human bookkeepers. Similarly software for scientists, engineers and decision-makers encourages calculations which would simply not have been considered prior to the invention of digital computers.

Another information-manipulation activity which has been enhanced by microbinic advances is text manipulation. Since text manipulation was transferred to a computer or dedicated wordprocessor, most routine manipulation tasks such as "cut and paste" and spelling checking have been programmed to be performed with a keystroke or mouse click. In addition, with an attached hard disk software procedures were invented to automate the filing and retrieval of documents. As text processing advanced specialized applications were invented. For example, today legal software with the capacity to rapidly search through text enables lawyers to quickly locate the precedents for a case in legal databases such as Lexis. In general, the invention of text manipulation software has greatly reduced the time and labor effort to make routine manipulations.

Once a text or numeric manipulation activity has been immersed in a microbinic medium economic incentives propel the development of software. For software to have value, the knowledge incorporated into software need not be a thorough understanding of the task but can be only surface knowledge, for example the intuitive opinions of an expert. Recently artificial intelligence inventors have created knowledge-based systems, which are frequently called expert systems.[62] Such software solves problems by mimicking the qualitative reasoning process an expert employs in solving a problem. Currently there has been reasonable success in the development and use of expert programs applied to problems with limited scope where the type of reasoning can be expressed as conditional judgments.

Currently the best knowledge-based systems can perform at the level of an intelligent assistant who can solve routine problems[63]. One successful application of such knowledge is software for the maintenance of physical equipment[64]. Another is a program called XCON, which is used by Digital Corporation sales personnel to configure VAXs. Even a profession as highly skilled as the medical profession has several expert program implementations which have yielded varying success. The program Mycin is better at diagnosing blood disorders than is the average doctor without specialized training. The economic value of software assistants stems from the fact that they are cheaper to create and operate than to train and employ a large number of human assistants.

An important advance enhancing the performance of individuals using all the types of software discussed above has been the movement towards developing integrated software. At first software programs for the various paperwork tasks performed on a desktop computer were incompatible and the user incurring a labor cost to transfer an information object from one program to another. Integrated software automates the transfer of information objects from one software application to another. An early example of this type of automation is the transfer of a spreadsheet from its data manipulation program to a wordprocessor program for inclusion in a document. Such automation requires the development of standard representations of information objects and the creation of software to translate information objects among competing standards.

As the power of desktop computers has expanded, software integration has also expanded to include all types of information objects. The most publicized example of this expanded integration has been the development of desktop publishing software, which integrates still image, data, and text manipulation activities in preparing documents[65]. Such software automates numerous tasks in editing and integrating test, data, and images in documents. One example of sound manipulation activities being integrated in personal computers has been the linkage of digital synthesizers to personal computers using the MIDI standard for music representation[66]. The current frontier is the creation of more powerful desktop computers and software to integrate dynamic image manipulation with the other types of information object manipulation to create a new form of communication, multimedia.

Advances in software and hardware create numerous opportunities for innovation which frequently require a complete reorganization of paperwork. Consider, for example, administration[67]. In the precomputer era, large institutions would organize large volumes of paper work as an assemblyline. This was the case, for example, with insurance claims. To achieve efficiency under such a system, people were specialized into filling one box on a form or performing a small number of routine operations. Later, in the era of standalone mainframe computers the paperwork assemblyline created the input for the mainframe. With the advance of dumb terminals and programming, the entire process of administrative activity is programed. For example, processing an insurance claim is contained within a computer program wherein a single operator could supply all the necessary information with prompts.

This reorganization constitutes an innovation in administration. A business can be more service-oriented in that a single service person can operate all the programs associated with a single client. And the programs can be more flexible than the human paper processing assembly line could be, thus enabling the organization to offer a greater variety of services. The single operator is in a much better position to correct mistakes than were the large number of people in the human assemblyline paperprocessing. This innovation has been imitated by public institutions such as universities where various operations on student records have been programed.

An amusing consequence of the advance of software and standalone hardware has been a large increase in the use of paper[68]. This is because software which increases the productivity in using standalone equipment, such as personal computers, also tends to create more output, which is usually in the form of paper. And at the same time, technological advances such as the zerox machine and the laser printer have increased the efficiency of paper manipulation tasks. Decreasing the use of paper depends on the development of communication protocols, software and networks in order to communicate information objects between standalone hardware using microbinic technology.

Nevertheless, it is precisely the rapid increase in the use of microbinics for information object manipulation that is creating powerful incentives for the expansion of microbinic communication. Performing a task requiring the use of a sequence of standalone hardware is inherently inefficient if the information objects must be converted to paper, or even tape, for transport between equipment. While some aspects of paper handling, such as letter sorting, can be automated, delivering the mail through microbinic communications channels facilitates automating the entire delivery process.

To transfer communication from paper to microbinics requires the adoption of standards for representing communication objects, communication protocols, communication networks with sufficient capacity and software to operate the communication system automatically. Because of the need to reach agreements on standards and the technical problems involved the advance of microbinic communication is taking place in three stages.

The first stage is microbinic communication among equipment within an office or order single site. This is because the institutions involved can purchase compatible equipment and software from vendors and establish a digital protocol for the office phone system or install a much higher capacity fiber optic network. Currently, a rapidly expanding technology is the development of local area networks to link mainframes, personal computers, mass storage devices, and other office equipment such as laser printers[69]. Apple Corporation's latest operation system for the Macintosh computer makes the transfer of information objects between application software of Macintosh computers in a Macintosh network automatic. Also IBM is currently promoting its Systems Application Architecture to promote software integration in networks[70].

Two economic factors are promoting this shift of internal corporation communications to microbinics. First, this linkage permits office personnel to share expensive resources effectively. Second, this transfer to microbinics facilitates the programming of many tasks in the filing and transmission of documents. Some corporations are now using optical scanners to convert all mail to microbinics when it enters the office. This step eliminates paper for intraoffice communication and displaces the file clerk and corporate postman using a shopping basket to deliver documents. Programing tasks in the newspaper business has linked the creation of the news with typesetting: software now takes news stories from the reporters' wordprocessors and automatically typesets the news story for printing.

The second stage of the development of microbinic communication is among offices and other sites within an institution. Again because a single institution is establishing the microbinic communication network, the problem of compatible equipment and software is easily overcome. What currently limits the growth of such communication is the limited capacity of the current analog-digital phone system and the expense of leasing higher capacity dedicated lines. Nevertheless, corporations are also installing corporation-wide electronic mail systems linking offices around the globe[71]. These local area networks and corporate-wide electronic mail systems are shifting the transmission of documents from paper to an microbinic medium.

Also data flows within the firm-even those with physically dispersed locations are being linked. Consider, for instance, the modern grocery store chain[72]. An optical reader at the checkout stand determines the product code. Once this is done the computer takes over by finding the current price and totaling up the bill. This programing of the task of determining the price eliminates the labor of stamping products with price data and the need for the checkout clerk to key in the price. This transferring the activity of price determination to microbinics also facilitates the programing of many other tasks such as the task of monitoring inventory levels. Linking data flows with the corporate office facilitates the automatic reordering of inventory and making detailed profitability calculations. For example, with the organization of data made possible by use of bar codes and optical readers, the supermarket computer can be programmed to compute the profitability of every square inch of display space and also to analyze the effectiveness of various types of promotions[73].

The goal of creating communication links between offices in an institution is the creation of a wide area network such that all the local area networks are linked into one giant network with the communication capacity of original local area networks. This is an area of active research and development currently[74]. Full realization of such a goal will probably be delayed until the phone system is based on optical fiber and the ordinary phone line has an enormous increase in capacity.

Advances in the manipulation and communication of information objects are creating a new framework for decisions in firms and other institutions. Consider the management of a corporation[75]. With the computerization of corporate records, a large database of corporate facts is created. The automation of the production process adds to this database a much better flow of production information than was previously possible to achieve. With increasing computerization of measurements, moreover, decision makers will have larger and larger databases of production, accounting, marketing and other data. And with the linkage of all offices and sites to the corporate mainframes, this data becomes increasingly available to corporate managers who analyse the data to make decisions. With new analytic tools such as data manipulation languages, spreadsheets, linear programming, and statistical packages, the databases support decisions based on analysis of the alternatives.

A consequence of this new framework for decisions is an ongoing reorganization of management hierarchies in corporations. The function of middle managers has traditionally been to filter the flow of information in passing reports up the chain of command. Currently executive assistants using the corporate decision support system can prepare such reports for the top executives without these intermediaries. Moreover, as information manipulation tasks become software intensive, these executive assistants can analyze a much greater variety of alternatives. Consequently, over time senior executives are reducing the number of levels in corporation hierarchies

The third stage in the development of microbinic communication is the development of interinstitution microbinic communication. Two impediments to such progress are the lack of standards and the low capacity of the existing phone system. The communication of text and data between institutions via microbinics has just begun. One important example is electronic mail, E-mail, a system which makes messages such as letters much less expensive to deliver electronically than physically, in the form of paper[76]. The major problem in the growth of electronic mail between institutions, however, was the lack of an universally agreed-upon standard for electronic mail. Thus, customers of the competing E-mail services could not communicate. With the adoption in 1989 of the international standard, X.400, which was created in 1984, all E-mail customers of all the E-mail services will be able to communicate with one another. The growth of E-mail should accelerate during the 1990s.

Market transactions are an important example of an interinstitutional paperwork activity whose productivity is being advanced by technological advances discussed in this section. To view this advance consider first the four components of market transactions: consideration of the alternatives, consummation of the transaction, payment, and finally the delivery of the good or performance of the service. As information manipulation in market activities is transferred to microbinics, the physical location of markets in which these steps are performed will become immaterial.

The types of markets which have progressed the furthest towards total transfer to microbinics are asset markets. The over-the-counter market for stocks, the Nasdaq, for instance, resides completely in a computer network[77] and the New York stock exchange is automating to provide the capacity of trading one billion shares a day[78]. To facilitate stock transfers, the stock certificate has become optional, and ownership is simply recorded in computer memory. Dow Jones and other services provide extensive information services to create databases for analysis of desired options. In addition, with brokers in information utilities such as Source, individuals can buy and sell assets twenty four hours a day

Progress towards transfer of market transactions to microbinics varies greatly between industries. In most markets this progress is based on the creation of databases that enable individuals to analyze alternatives and in some cases consummate transactions. Perhaps the most highly developed example occurs airline reservation systems[79]. These systems are very large databases maintaining information about the status of seats reserved for all flights covered by the system. The database is available at terminals in airports, travel offices, corporations, and in the home through information utilities. Reservations are made by interacting with the system. Numerous tasks such as printing tickets, assigning seats, and searching for the lowest price have been programmed. Other less developed databases exist for other markets such as real estate.

A specialized type of market transaction which is currently being transferred to microbinics and programmed is ordering routine purchases such as parts from suppliers and inventory for retail stores. In making routine purchases it is much more efficient for the computers of the respective firms to communicate directly with one another, thereby eliminating the intermediate step of transfer to paper. But to eliminate this intermediate paper step, conventions must be established so that the computer and software of the buyer can communicate with the computer and software of the supplier. Currently these conventions have been established in the ANSI X.12 standard whose acceptance is growing among suppliers and buyers by the prospect of reduced paperwork costs[80]. In programing the ordering of parts from suppliers, for example, the big three auto makers expect to save some two billion dollars a year[81].

Household market transactions are just beginning to be transferred to microbinics. Systems for paying bills by telephone as well as telephone query systems are in operation. More widespread merchandise sales, such as those which have been carried out through sales catalogs ever since the 19th century, are just now being transferred to microbinics. As womens' participation in the workforce has increased, catalog purchases have become a growing portion of consumer purchases. Beside the more traditional mail and phone orders for which the buyer uses a printed catalog to make comparisons, two new forms of markets in communications networks have been created. One of these is computer stores found in bulletin boards and information utilities such as Compuserve; another is the new version of home shopping that has been introduced on cable television. This television market has the quality of a game show, with the viewer having only one option, to either buy or not buy the merchandise being displayed, using a phone to complete the transaction. As the telephone and television shift to digital, these growing services will be transferred to microbinics. In all markets, moreover, the payment mechanism is being transferred to microbinics in a piecemeal fashion. Currently, payments for goods and services are made primarily with currency, checks and credit cards. Also a new type of card called a debit card is gaining acceptance in obtaining cash and making deposits at automatic bank tellers. The use of debit cards is just beginning to compete with the use of credit cards is making purchases. The use of a debit card to make purchases totally automates the payment mechanism because funds are instantaneously transferred from the customer's account to the store's account.

Programing is automating various tasks in the use of all types of payment mechanisms. In check clearing, the account identification numbers are inputed into the computer automatically, limiting the human input to the amount on the check. As the practice of not returning the canceled checks gains universal acceptance, all subsequent operations to clear the check can be handled automatically[82]. Validation of credit cards in stores is another example of how the payment process is now automatic. Obviously, however, while advances are being made with various component operations, the current media of market transactions is a mix of paper and electronics.

Notes and References 3

1. Gellhorn, W., C. Byse, and P. L. Strauss, 1979, *Administrative Law*, (The Foundation Press: Mineola)

2. Wilson, W., 1913, *The New Freedom*, (Doubleday and Company: New York)

3. Inose, H. and J. Pierce, 1984, *Information Technology and Civilization*, (W. H. Freeman and Company, New York)

4. Nash, J., 1990, A bumper crop of biotech: genetic engineering promises to transform agriculture. *Time*, Oct 1, pp 92-93; Hamilton, J., 1992, Biotech: America's Dream Machine *Business Week*, Mar 2, pp 66-74. It is possible that biotechnology may suffer the same fate as expert systems in that substantial progress may take much more effort than current media hype suggests.

5. Becker, J. D., 1984, Multilingual Word Processing, *Scientific American*, Jul, pp 96-107

6. Wolfram, S., 1984, Computer Software in Science and Mathematics, *Scientific American*, Sep, pp 188-204

7. Shell, E. R., 1981, Bach in Bits, *Technology Illustrated*, Oct/Nov,pp 21-29

8. Monforte, J., 1984, The Digital Reproduction of Sound, *Scientific American*, Dec, pp 78-84

9. Bostelmann, G. and P. Pirsch, 1985, Coding of Video Signals, *Electrical Communication*, Vol 59, No3.

10. Staff Conrac Division, Conrac Corporation, 1985, *Raster Graphics Handbook*, (Van Nostrand Reinhold Company: New York)

11. The progression to greater resolution in computer displays is well known to any reader who has followed the evolution of the graphics standards for the IBM PC and clones. At the same time much progress has been made in creating flat screen technologies. See: Woodard, O. C. Sr. and T. Long, 1992, Display Technologies, Byte, Jul, pp 159-168. For flat screen technologies to totally replace CRTs they will have to gain both a performance and cost advantage.

12. Roth, C. H.,1979, *Fundamentals of Logic Design, 2nd Edition*, (West Publishing: St. Paul)

13. Although now dated, an excellent survey of microelectronics is the September 1977 issue of *Scientific American*

14. Noyce, R. N., 1977, Microelectronics, *Scientific American*, Sep, pp62-69

15. Procassini, A. A., 1989, An Overview of the U.S. Semiconductor Industry, Seminar presented the Semiconductor Trade and Competitiveness Symposium, University of Texas, 3 Mar 89.

16. In this area the Japanese have displaced the Americans as the principal producer and have sparked a major reconsideration of American trade policy. See: Prestowitz Jr., C. V., 1988, *Trading Places*, (Basic Books, New York)

17. Trimberger, S. and J. Rowson, 1987, CAD for Building Chips, *BYTE*, June, pp 217-224

18. See 14

19. Meindl, J. D., 1987, Chips for Advanced Computing, *Scientific American*, Oct, pp 78-90

20. See 19

21. Matisoo, J. 1980, The Superconducting Computer, *Scientific American*, May, pp 50-65. This technology is not without its problems as the following article attests: Staff, 1983, Why IBM Gave Up on the Josephson Junction, *Business Week*, Nov 21, pp 76

22. See the In Depth report on optical technologies in the Oct 1989 issues of *Byte*. Another possiblity is integrated circuits based on the movement of a single electron. See Likharev, K. and T. Claeson, 1992, Single Electronics, *Scientific American*, Jun, pp 80-85.

23. Toong, H. D., 1977, Microprocessors, *Scientific American*, Sep, pp 146-161

24. Brand, R, 1987, Chipmakers Are Taking a Gamble on RISC, *Business Week* July 20, pp 104-105 Since 1987, SUN has promoted the RISC based workstation to become the leader in the market.

25. For a forecast of advances in computing and software see the October 1987 issue of *Scientific American*

26. For a brief overview of the design of a personal computer see: Toong, H. D. and A. Gupta, Personal Computers, 1983, *Scientific American*, Dec, pp 87-106

27. For example, see: Hills, W. D., 1987, The Connection Machine, *Scientific American*, Jun, pp 108-115; Davis, D. B., 1987, Parallel Computers Diverge, *High Technology*, Feb, pp 16-22; and Byan, B., 1991, Multiprocessor Surf's Up, *Byte*, Jun, pp 199-206

28. O'Reilly, B., 1989, Computers That Think Like People, *Fortune*, Feb 27, pp 90-93

29. For a survey of software issues see the Sep 1984 issue of *Scientific American*

30. Tesler, L. G., 1984, Programming Languages, *Scientific American*, Sep, pp 70-93 The current rage in programing languages is object-oriented programing. This concept has the long term potential of transforming the creation of application software into the assembly of components. See: Haavind, R., 1992, Software's New Object Lession, *Technology Review*, Feb/Mar, pp 60-66.

31. Foley, J. D., 1987, Interfaces for Advanced Computing, *Scientific American*, Oct, pp 127-135

32. For a nontechnical account of communication, see: Pierce, J. R., 1981, *Signals*, (W. H. Freeman and Company: San Francisco)

33. Heldman, R. K., 1988, *ISDN in the Information Marketplace*, (TAB Professional and Reference Books: Blue Ridge Summit, PA)

ISDN has been adopted in Europe and France leads in ISDN applications. Since ISDN was formulated advances in communication have demonstrated that ISDN has inadequacies and therefore it is possible that ISDN may never become widely accepted in the US. Nevertheless, by the end of 1991 the communication industry had agreed on the ISDN 1 standard for ISDN implementation which indicates considerable progress. For the purposes of this book it is immaterial whether the communication industry adopts ISDN or a newer improved digital protocol. The book is based on a forecast that the communication industry will first adopt a narrowband standard for digital communication over copper wire and then expand the standard to include broadband communication over optical fiber. For the purpose of discussion the author will assume ISDN will be widely adopted.

34. Shannon, C. and Weaver, W. 1949, *The Mathematical Theory of Communication*, (The University of Illinois Press, Champaign). See 32 for a non-technical discussion

35. For a survey of the 1992 status of image compression see the special supplement: Image Compression of the Mar/Apr 1992 issue of *Computer Pictures*

36. Sanferrare, R. J., 1987, Terrestrial Lightwave Systems, *AT&T Technical Journal*, Vol 66, Issue 1, pp95-107 and Desurvire, E., 1992, Lightwave Communications: The Fifth Generation, *Scientific American*, Jan, pp 114-121

37. One of the first books on automation is: Diebold, J., 1952, *Automation*, (D. Van Nostrand Company,Inc, New York). The September 1982 issue of *Scientific American* is devoted to automation.

38. For a more detailed breakdown see: Gunn, T., 1982, The mechanization of design and manufacturing, *Scientific American*, September, 114-131

39. For example see: Computer-Aided Design, *Byte*, June 1987, pp 177-224; Klee, K., 1982, CAD/CAM (5 articles), *Datamation*, February, pp 110-144; Hordeski, M.F., 1986, *CAD/CAM Techniques* Reston Publishing Co: Reston); Lerro, J. P., 1982, CAD/CAM system, *Design News*, 3-1, 34-44; Teresko, J., 1983, CAD/CAM goes to work, *Industry Week*, Feb 7, 40-47; Krouse, J. K., 1984, Automation Revolutionizes Mechanical Design, *High Technology* Mar pp 36-45; Smith, M. R.,1986, Automating the Styling, *Automotive Engineering*, Mar, pp31-35; Francett, B., 1988, Turning Visions into reality, *Computer Decisions*, Oct, pp 28-29. As hardware and software advance and hardware costs fall, CAD programs become ever more powerful. For example, current CAD software for architecture allows designers and client to walk through buildings before construction. See: Greenberg, D, 1991, Computers and Architecture, *Scientific American*, Feb, pp 104-109

40. Wright, J., 1987, Intelligent CAD, *Systems International*, Apr, pp 43-44

41 For the design of automatic design of circuits for component lists see: Crooke, A. W., 1987, The CADcompiler, *Byte*, June, 187-197. Software to design an integrated circuit is known as a silicon compiler, for example, see reference 17.

42. Krouse, J. K., 1986, Engineering Without Paper, *High Technology*, Mar or more technical see: Ohr, S. A. 1990, *CAE: A survey of standards, trends, and tools* (Wiley: New York)

43. Cook, N. H., 1975, Computer-managed Parts Manufacture, *Scientific American*, Feb, pp 22-29

44. Zygmont, J., 1986, Flexible Manufacturing Systems, *High Technology*, pp 22-27 or Greenwood, N.R., 1988, *Implementing Flexible Manufacturing Systems* (John Wiley & Sons: New York)

45. Bylinsky, G., 1983, The Race to the Automatic Factory, *Fortune*, Feb

21, pp52-64

46. Mtichell, R., 1986, Automation is not Just for the Big Guys Any More, *Business Week*, Jan 27, pp 96H

47. Sepehri, M., 1985, How Kanban System is used in an American Toyota Motor Facility, *IE*, Feb, pp 50-56. or see: Dear, A. 1988, *Working Towards JUST-IN-TIME* (Van Nostrand Reinhold: New York)

48. Otis, P, 1987, Quality, *Business Week*, Jun 8, pp 130143

49. Leffler, A. L., 1989, Material Handling–The New American Manufacturing, *Automation*, Feb, pp 20-22

50. Durkee, D., 1984, MacFactory!, *Softtalk*, pp 130-136

51. Horn, B.K.P. and K. Ikeuchi, 1984, The Mechanical Manipulation of Randomly Oriented Parts, *Scientific American*, pp100-111

52. Seering, W. P., 1985, Who Said Robots Should Work Like People?, *Technological Review*, Apr, pp 59-67

53. Shapiro, S., 1986, Electronic Assembly Becoming Dependent on Robotic Tools, *Computer Design*, Feb 1, pp 33- 36

54. Stovicek, D., 1988, Machines That Keep an Eye on Production, *Automation*, Jul, pp 12-16

55. Bylinsky, G., 1986, A Breakthrough in Automating the Assembly Line, *Fortune*, May 26, pp 64-65

56 Koenig, D. T., 1990, *Computer Integrated Manufacturing: Theory and Practice* (Hemisphere Publishing Corp: New York)

57. Mayer, R. J., 1987, IGES, *Byte* Jun, pp 209-216

58. Bairstow, J. 1986, GM's Automation Protocol, *High Technology*, Oct, pp 38-42 or see for a technical account: Pimentel, Juan R., 1990 *Communication Networks for Manufacturing* (Prentice Hall: Englewood Cliffs). MAP has its problems. See: Stix, Gary, 1991, Off the MAP, *Scientific American*, Aug ¡ pp 100,101.

59. Brody, H., 1987, CAD Meets CAM, *High Technology*, May, pp 12-18

60. MicroCAD New Staff, 1988, Pairing CAD and CAM Systems for Manufacturing, *MicroCAD News*, Sep/Oct, pp 31-35. A new industry called desktop manufacturing is emerging for making prototypes for manufacturing development. In desktop manufacturing design is completely integrated with part fabrication. The current limitations are the types of materials out of which parts can be fabricated. See: Stix, Gary, 1992, Desktop Artisans, *Scientific American*, Apr, 141,142

61. Saporito, Bill, 1986, IBM's No-hands Assembly Line, *Fortune*, Sep 15, pp 105-107

62. Winston, P.H., 1984, *Artificial Intelligence*, (Addison-Wesley: Reading)

63. Dreyfus, H. & S., 1986, Why Computers May Never Think Like People, *Technology Review*, Jan, pp 43-61

64. For a list of numerous working systems see: Buchanan, B. G., 1986, Expert Systems: Working Systems and the Research Literature, *Expert Systems*, Jan, pp 32-51

65. For example, see: Seybold, John W., 1987, The DesktopPublishing Phenomenon, *Byte*, May, pp 149-154 or Hammonds, K, 1988, These Desktops Are Rewriting the Book on Publishing, *Business Week*, Nov 28, pp 154-156

66. Boulez, P. and A. Gerzso, 1988, Computers in Music, *Scientific American*, Apr, pp 44-50 and Thompson, T., C. Wolff and D. Cook, 1987, Music Is Alive with the Sound of High Tech, *Business Week*, Oct 26, pp 113-116

67. Giuliano, V. E., 1982, The Mechanization of Office Work, *Scientific American*, Sep, pp 149-165

68. Tenner, E, 1988, The Paradoxical Proliferation of Paper, *Harvard Magazine*, Mar/Apr, pp23-26

69. Lefkon, D. 1987, A LAN primer, *Byte*, Jul, pp 147-154

70. Verity, J. W., 1989, A Bold Move in Mainframes, *Business Week*, May 29, pp 72-78

71. Hamer, M. and J. Heilmann, 1988, How One Firm Created its Own Global Electronic Mail Network, *Data Communications*, Jun, pp 167-184

72. Staff, 1981, Supermarket Scanners Get Smarter, *Business Week*, Aug 17, pp 88-92

73. Taylor, T. C., 1986, The Great Scanner Face-Off, *S&MM*, Sep, pp 43-46

74. Tazelaar, J., R. Green, B. Ryan, J. Barron, P. Stephenson, S. Fisher, and A. Gore, 1991, State of the Art–Wide-Area Networking, *Byte* Jul, pp 158-189

75. Staff, 1983, A New Era for Management, *Business Week*, Apr 25, pp 50-86

76. Rothfeder, J., 1989, Neither Rain, nor Sleet, nor Computer Glitches, *Business Week*, May 8, pp 135-139

77. Louis, A. M., 1984, The Stock Market of the Future-Now, *Fortune*, Oct 29,

78. Layne, R., 1988, Exchange Readies for Billion-Share Day, *InformationWEEK*, Oct 3, pp 20-21

79. Dunn, Si, 1984, Super SABRE, *American Way*, Sep, pp 76-83

80. Kelleher, J., 1986, Electronic Data Interchange, *Computerworld*, Sep 22, Worldwide use of EDI to reduce paperwork in transactions will accelerate as communication standards are accepted. See: Kukmar, A. 1991, EDP and Message Handling Systems, *C&C*, Mar, pp 53-58

81. Mitchell, R., 1985, Detroit Tries to Level a Mountain of Paperwork, *Business Week*, Aug 26, pp 94-96

82. Ernst, M. L., 1982, The Mechanization of Commerce, *Scientific American*, Sep, pp 132-147

Chapter 4

Forecast

Introduction

From the time Karl Marx made his gloomy forecast of the dynamics of capitalism in *Das Kapital*[1], the specter of massive displacement of the workforce by technology has been part of intellectual history. From its beginnings in the 18th century until very recently, the process of applying technology to production could be described as mechanization. In most mechanized production processes man acted as a intelligent, flexible control device for dumb, powerful, and inflexible machines. In some mass production plants mechanization led to hard automation in which human skills were incorporated into the machinery. Except for sporadic depressions, massive unemployment never materialized as mechanization advanced, since job creation in the labor intensive services was much greater than job destruction from mechanization in both agriculture and manufacturing. With a continual rise in productivity real wages increased accordingly.

Because job creation in the post World War II era has been great enough to absorb not only displaced workers but also the additional employment demands created by the baby boom and women entering the workforce, why should the new soft automation, in which human knowledge and skills are incorporated into software, cause any concern in the future? Soft automation raises a major concern because soft automation greatly expands the scope of tasks which can be automated. The flexibility which soft automation makes possible means that soft automation can be economically applied to batch as well as mass production. Second the new soft automation can be applied to the manipulation of information objects as well as physical. This means that soft automation can be applied to services as well as manufacturing.

Whether advancing automation promotes political stability depends on the resulting income distribution. With a high rate of investment in the economy, the total output of goods and services will increase, indicating an increasing per capita GNP. Empirically, increasing real income for nearly all members of society promotes political stability. With soft automation it is far from obvious that economic forces will generate a politically stable income distribution. An important issue is the long term impact of automation on the income distribution. For the rest of the book the word automation will imply soft automation unless specifically referred to as hard automation.

Development of the Social Nervous System

As was pointed out in the previous chapter, advances in automation require the control of the activity be transferred to microbinics. For example, the automation of tasks involving the physical manipulation of objects requires the development of a computer model of the process. Also, computer integrated manufacturing requires communication among all factory machines. Programing tasks in information manipulation is promoted by transferring the activity to microbinics. Advances in computation and communication will create a social nervous system which will provide a unified basis for the manipulation, storage, and communication of all types of information objects. The creation of this social nervous system will, in turn, promote great advances in automation.

The rate of expansion in computer power should continue at current rates until well into the next century before gradually slowing down[2]. While the rate of increase in the number of components in an integrated circuit will gradually slow, increases in computing power will continue as inventors shift their attention from single processors to parallel processors and to new technology such as optical computers. To exploit the potential power of parallel processors, software developers will create numerous new parallel processing languages. Thus, every eight years or so the power of a particular type of computer, for example, a personal computer, will increase by a factor of ten. At the same time, advances in memory devices such as optical disks will enable computers to access prodigious amounts of information such as entire libraries.

Besides simply making advances in the speed of executing instructions, developments in computing will expand the focus of the computer beyond the current focus on data and text processing. Inventors will make significant advances in pattern recognition capability through advances in neural networks. A considerable portion of the increased computer resources will be devoted to better graphics and at the same time computer screens will be able to display finer and finer detail in ever more shades of colors. The great increase in computer power will make the manipulation of sound and all types of static and dynamic images as easy to perform as current manipulations of numbers and text.

Advances in programming will increase the productivity of both programmers creating software and the workers using software. High level programming languages will specialize to provide application programmers in each type of application with intuitively obvious instruction sets. One example is "English" type commands for manipulating a database. Language developers will incorporate artificial intelligence aids to perform routine operations automatically. However, as discoveries in the understanding of the more creative aspects of programming are not likely to lead to accurate mathematical models any time soon, the complete automation of programming is unlikely in the foreseeable future. From the perspective of the end user, computers will become increasingly "user friendly." Voice activated languages will compliment visual icons to make using computers intuitively easy for all. Because of falling cost and better performance, every individual at both work and in the home will have access to a computer.

The expansion of the capacity of the communication system will take place in two stages. Currently the local phone service is analog which when used with the currently most popular modem transmits 2400 bits/second. As the need arises, phone customers can lease higher capacity analog and digital lines. The first stage in the expansion of communication capacity is presently focused on extracting more capacity out of the twisted, copper wire telephone system with the shift from analog to digital under the worldwide ISDN standard[3]. The basic phone service for the ISDN standard is 144,000 bits/ second partitioned into three channels. Again as the need arises, customers will be able to lease higher capacity digital lines. The second stage in the expansion in the communication system is the replacement of the twisted wire local communication system with optical fiber and the development of high capacity switches. With the shift to optical fiber the basic phone service will probably be defined in the range of 150 to 160 million bits/second under a broadband ISDN standard[4].

The shift will be transparent to the user, who will maintain his analog phone until his demand for an increasing array of digital communication services prompts him to switch to a higher capacity digital line. The wire ISDN standard has been in use in Brittany, France since 1987[5] and is being tested for use in the US. The shift to ISDN will be slow at first because there are few ISDN services and the cost of equipment is high because of small production lots. As services using the ISDN are created and the cost of ISDN equipment falls due to much larger production runs, the use of the wire ISDN will quickly expand. Large firms and government will shift first, followed later by small firms and households. While many activities are now partially transferred to microbinics the integrated wire ISDN communication system will spur a more complete transfer.

The shift to the high capacity fiber optics for local communication will start prior to the year 2000 and will follow a similar pattern as the shift to the wire ISDN standard. Thus, long before the shift to the basic wire ISDN standard is complete, institutions with large demands for communications will be shifting to fiber optics. In the 1990s as the cost of installing fiber declines to the level of installing copper, increasingly new phone service will employ fiber in anticipation of future growth in communication demand. Phone companies have already replaced copper wire by fiber for high capacity trunk lines. In the next thirty years phone companies will be replacing all twisted wire lines with optical fiber. The use of fiber optics for local communication will spur the shift of most information manipulation activities involving text, data, symbols, voice, and all type of images to integrated computer-communication networks.

The expanding capability of computing, storage, and communications is only one factor in the shift of activities to microbinics. The major hurdle to this transfer is, perhaps surprisingly, the problem of creating universally agreed standards for representing the information objects of various activities. Reliance on government to accelerate the creation of these standards will not necessarily improve the situation. In fact, competition between rival groups frequently results in the creation of better standards than those which might be set too quickly or arbitrarily. Once standards for a particular type of information object have been created, the manipulation and communication of these objects shifts to microbinics.

Consider first the shift of most activities involving text to microbinics. Two technical factors are currently promoting the shift of text and documents within institutions to microbinics. First, the technical problems of creating local area networks to link equipment within institutions have currently been solved, although a great deal of work needs to be done to integrate rival standards. Second, mass storage devices such as optical readers allow institutions to store massive amounts of text and other documents and to treat text as an image. The progress towards the creation of a paperless factory by General Motors and a paperless office by the Pentagon[6], will create standards and provide useful experience for other institutions to follow.

Currently, text written on paper is easier to read than text written on computer screens. In about ten years, however, the quality of computer screens will improve so that reading screens will be easier. This will accelerate the complete transfer of paperwork within institutions from paper to microbinics. The complete transfer will require some type of device, which inhibits tampering such as a permanent one-time-write memory device, as a legal means of record[7].

Currently, the communication of text between institutions via electronic mail is just now beginning to accelerate. The adoption of the international standard, X.400, has provided a basis for linking all senders once the electronic mail directory is complete in 1993[8]. As this standard becomes universal, vendors of local area networks to promote sales will have to ensure that their network can interface with the X.400 standard. Once this has been accomplished the volume of electronic mail will explode because electronic mail is both much faster and much cheaper than paper mail.

In a similar fashion most activities involving data will also gradually be completely transferred to microbinics. As decision making becomes increasingly analytic, based on more detailed models, the flow of data will increase. Today, an important growth area in data communications is among the physically dispersed operations of a firm. As standards for internal communications within institutions are created, the transfer will be complete. The creation of standards such as those proposed by IBM in their system applications architecture will promote interinstitutional data communication.

Two important incentives for shifting information manipulation activities to microbinics are cost reduction and speed. It is much less costly and much faster to communicate the binary representation of an information object than to transport the same representation on paper. Moreover, a complete shift to microbinics generates much greater savings. Currently information objects are transferred back and forth from paper to microbinics in activities such as ordering supplies. A complete transfer to microbinics also eliminates the labor costs involved in these transfers, and would eliminate the need for two communication systems, one paper and the other microbinic.

The cost reduction in shifting activities to microbinics can be considerable. For example, in the case of ordering parts for automobiles the shift is expected to save as much as $200 per automobile[9]. Because of the projected savings, the transfer to microbinics of the ordering of routine purchases, such as inventory by department stores, which is just being initiated, should be moving towards completion within a decade.

The increased capacity of the communication system and the creation of standards will greatly increase the shift of various market transaction activities to microbinics. An example is the creation of a real-time payment mechanism or point-of-sale, POS, system. The debit card which currently is used to obtain cash from automatic teller machines is an obvious candidate for this role in retail purchases. Software already exists for interfacing the automatic teller machines with various banks. Numerous pilot POS systems are in operation. The primary hurdle to create a complete POS system is the need for the various participant: merchants, banks, and vendors to reach agreement on the standards for the system[10]. For example, should the POS system work instantaneously or overnight? The system design must also be flexible to meet the needs of the various types of transactions to be processed. The problem with creating a fast, reliable, inexpensive real-time payment mechanism is that there is no simple market mechanism for creating the transformation. Indeed, the public will be slow to accept such a mechanism.

Automating the payment mechanism between economic agents is a public good because once the total transformation takes place, all will benefit through greater efficiency. Stores will find automatic payment advantages, in that bad checks will be eliminated and credit charges will be paid without delay. Once volume has been established,the cost of clearing transactions microbinically should fall to less than one tenth the cost of clearing a check. While consumers would lose the float or the time delay between the time when items are charged and the time when they are paid for, the loss of these advantages would be compensated by lower prices and a wage-and-salary-payment system much closer to real time, based for example, on daily or even hourly intervals.

As activities shift to microbinics, the demand for digital communications provided by the ISDN standard will grow because firms can reduce costs by shifting the respective office or market activity to microbinics. Thus the initial demand for ISDN standard is the business demand to transmit data, text, and FAX. But as higher capacity lines come into use entrepreneurs have strong incentives to create additional communication services. For example, businesses will want to link local area networks at physically dispersed locations into one giant local area network. Communication carriers will respond to increased demand by increasing supply.

Entrepreneurs, responding to opportunities created by the expanded communications channels under wire ISDN, will create many new services. These new services will involve shifting new activities to microbinics. Underlying the potential demand for new services is the fact that the communication of a bit of information is a decreasing cost industry, whereas travel, especially if one considers the value of ones time, is an increasing cost industry. One important factor in the growth of new services in communication networks will be the substitution of communication for travel. One component of the creation of these services will be the transfer of information object manipulation to microbinics.

One current growth industry in communication networks is the creation of databases. As the capacity of communication expands, the creation of ever more extensive information retrieval services increases. Consider, for example, a library. One fundamental problem in the current paper organization is that if a library only has one copy of a book, only one person can read the book at one time. If the library is moved into an electromagnetic environment, however, anyone who wants a copy can have one. Moreover, placing the library in an electromagnetic environment eliminates the time necessary for physically traveling to the library. Finally, a microbinic library on some mass storage device such as an optical disk occupies much less space than a physical library.

The complete shift of libraries to microbinics is taking place in stages. Currently many libraries have computerized the checkout of books and other administrative procedures. Software to search for material is becoming increasingly sophisticated. An example of search software is the Lexis system which enables lawyers to search for precedents to prepare a case. With the advance of optical disk technology an increasing amount of text is being transferred to microbinics[11]. At the current stage of development the primary access to text on optical disks such as CD-ROMs is through microcomputers. With the expansion of communication channel capacity via ISDN and the resolution of copyright concerns, books and other large amounts of texts will be transferred to users through communication channels.

The first group to use the basic wire ISDN service in the home will be millions of full and part time telecommuters, who will need this service to be effective in their jobs. A growing number of homes with the basic ISDN service will create a market for entrepreneurs to develop new information services for the home. One example will be the public library. Another example is much improved home shopping. As was pointed out in the previous chapter, with two-income families, home shopping, primarily through catalog sales, is growing faster than mall sales. Home shopping in communications networks has serious defects. Telemarketing on cable television is not interactive, hence the viewer can not search for a product he wants. Computer stores in information utilities such as Source provide only the model number as information concerning each product. Recently, Trintex, a IBM/Sears videotex venture created Prodigy, an improved shopping and information service through computer networks[12].

As the capacity of communications channels expand, the quality of home shopping services will improve dramatically. The digital expansion of the telephone system would greatly facilitate catalog sales through computers. The telephone system is interactive, hence the viewer seated at his terminal can search through the catalog on the screen at his leisure. The ISDN standard will create the capacity for the transfer of sufficient information to describe the product clearly. In a few seconds the viewer could obtain a specification page. Also, if he desired, the viewer could obtain a static image. The quality of the image depends on the length of time the viewer is willing to wait, and on the quality of compression algorithms used to reduce the required data transmission. When using the basic wire ISDN service to view a catalog, a picture would be an option with alternative quality levels and time delays.

While the increases in computer capacity will take place in uniform, discrete jumps, the potential increase in communication capacity from the basic wire ISDN standard to a basic fiber optic ISDN standard is a prodigious jump. With twisted wire as the basis for local communication, the phone system can not transmit dynamic video clearly without repeaters. If the basis for local communication becomes a fiber optic line, however, the potential capacity of the local network becomes thousands of television channels. To shift completely to fiber optics will require communication exchanges or switches capable of switching very high volume digital messages. These high volume switches should be created by 1995[13].

Several factors will promote the shift of local communications to fiber optics. Firms and other institutions will want such high capacity local communications for bulk data and text transfers both within and among institutions. However, the factor most likely to promote the shift to a complete fiber optics system will be the growth of teleconferencing. In arranging meetings among persons at physically dispersed locations, the advantage of a face-to-face meeting is the personal contact, while the disadvantages are the cost of travel, the time delay in organizing a meeting, and the less-than-optimal use of time while traveling. Currently interactive video teleconferencing is a rapidly growing, declining cost technology for group meetings. Because high quality, interactive video teleconferencing can achieve most of the advantages of a face-to-face meeting, institutions will demand high capacity, low cost fiber optic communications.

The declining costs of fiber optics will spur the growth of interactive, video teleconferencing even further. Teleconferencing will start as a substitution for meetings, but as its cost declines, teleconferencing will be used to create temporary groups within organizations without the expense of temporary relocations. As cost further declines, teleconferencing will become a mechanism for organizing groups at arbitrarily dispersed locations through intelligent terminals. Institutional reliance on teleconferencing as a means for organizing groups will depend on innovators who learn how to achieve high group performance in groups organized through the social nervous system through appropriate technology, software and incentives.

While institutions will desire fiber optics as a low cost means of organizing groups at dispersed locations, individual will desire such organization to achieve greater control in their daily schedules. For example, the minute an individual logs into his or her work group at an intelligent terminal, software will immediately link the individual to the appropriate network. Such an individual would be fully functional at any intelligent terminal; consequently, the individual could be at corporate headquarters, a remote corporate office, a neighborhood office or at home. Moreover, software will enable individual to appear to members of his or her group dressed appropriately for the corporate culture. That is the individual could appear dressed in a conservative business suit while seated in a swim suit at a terminal by a pool. Moreover, given the tremendous capacity of fiber optics, individuals besides their formal work networks will create private, possible encrypted networks for informal communication. Thus individuals would not need to feel isolated from their institution while working though the social nervous system.

A second aspect of the increasing demand for fiber optics as the basis for local communications will be the growth of interactive video services[14]. Interactive video technology will impact many activities such as education and marketing. Interactive videos can be stored on optical disks, a very large storage disk read by a laser stylus. Optical disk technology is just beginning to fall in price so that it is accessible to firms and wealthier households. With twisted copper wire communications, this technology is accessible to those who own optical storage devices and buy the disks. For disks which many individuals might wish to use for only a short time, there are enormous economies of scales if the disks could be addressed remotely through communication channels. Such communication would require optical fiber in order to be effective.

With large volume communications made possible by fiber optics, catalog sales would be enhanced with the option of video demonstrations of the product. With the introduction of cheap, interactive-image communication, the possibilities for much greater choice and better information support for comparison shopping will be realized. The advent of fiber optic communications in the household will enable the shopper to obtain as much visual information from a much better screen as he currently obtains from observing store displays. For example, inexpensive image technology will enable real estate agents to show images of alternative property along with data now found in computerized listing services. With fiber optic communication in the home, the growth areas of merchandising will be interactive image technology as an advance over paper catalog sales, computer stores and home shopping networks on television. As will be discussed in the next two chapters, economic incentives will create markets based on interactive image communications supported by extensive databases of facts and images for decisions. As the supply of image services in communication channels grows, fiber optics will enter households providing a single communication link absorbing all communications such as cable television.

Within the next 50 years, as the growth in communication services begins to create demands for a high capacity optical fiber basic service for local calls, the entire twisted copper-wire phone system will be replaced with a fiber optics system. A fiber optics line in every office and household would create a single communication system for all types of communication signals, whether one-way, such as cable television, or interactive, like voice. Having a network of optical fibers in place, moreover, would create a framework for increased use of the potential capacity as the demand for telecommunications, telecommuting and interactive video communication services increases over time.

What the advances in microbinics provide is a unified basis for manipulating and communicating words, numbers, equations, voice and images. With the simultaneous advances in microelectronics and communications, the two technologies are being fused. With the advance of microbinics, society is thus evolving a social nervous system, which integrates computer, communications, and all library collections. It follows that the transfer of activities to microbinics means that these activities are being transferred to the social nervous system. The creation, manipulation, and communication of all informational objects whether numbers, text, symbols, voice or image will increasingly take place in this social nervous system. This means that books, papers, film, conversations and so on would be stored and transmitted electromagnetically.

As the capacity of the communication system expands,the manipulation, storage, and communication of all types of information objects will become integrated in the personal computer, which will operate as an intelligent terminal in the social nervous system. One example of this integration is the creation of FAX cards for PCs. Another is the growth in the market for multimedia computers[15]. Yet another is the construction of an intelligent terminal by Datapoint which integrates voice, data, text, symbol, still image and poor quality video in one terminal. This terminal should be considered a prelude of future developments by many companies[16].

Viewing of informational materials in the social nervous system would generally be done at intelligent terminals using various types of screens or projectors. From each terminal each person would have potential access to all of the society's computer power, and the entire store of society's accumulated knowledge in the form of books, data files, video, and so on. Also, from each terminal each user would have potential communication with any other user. Throughout this book the word "terminal" will refer to a general-purpose device which can be used for voice-video communication and computation either as a smart terminal or as an input device for a larger computer in the social nervous system. A terminal may be used for word processor, or any other type of information object manipulation.

Automation: Physical Manipulation

The next step in estimating the impact of automation on the income distribution is to estimate the impact of automation on the total hours of employment per week. The total hours of employment per week is simply the number of workers times the average length of the work week. The trend in the average work week will be considered subsequently. In addition, the impact of automation on the skill requirements of tasks performed by humans will be considered.

To consider the impact of automation, the total hours of employment per week will be divided into blue-collar work and white-collar work. This division will be made using the Bureau of Labor Statistics occupational groupings[17]. Blue-collar work will consist of the total hours of employment per week in occupations under the headings of service operators; precision production, craft, and repair; operators, fabricators, and laborers; and farming, forestry, and fishing. Thus defined blue-collar work consists of 45% of all civilian work and primarily involves manipulation of physical objects. White-collar work will consist of the total hours of employment per week in occupations under the headings of managerial and professional specialty and technical, sales, and administrative support. Thus defined white-collar work consists of 55% of all work and involves primarily manipulation of information objects. The decline of blue-collar work will be considered first.

First, let us consider manufacturing-blue-collar work, which currently comprises less than 10% of all civilian work. The economic incentives for a firm to promote automation are much more than simply reducing labor costs. Another advantage of automation is the flexibility in producing batches of goods. With the capacity to shift rapidly from the production of one part to another, production can respond quickly to changes in the marketplace. The resulting reduction in setup times reduces the costs of batch production towards the level of mass production with fixed machines. And as the costs of automated production fall, large firms will increasingly switch mass production techniques from hard to soft automation in order to obtain greater flexibility in production.

Automation can also aid in quality control. Automation encourages product and production simplification. As Frost Inc discovered, installing a flexible manufacturing system can reduce the number of defective parts. The installation of an automatic delivery system for parts in the assembly process to implement just-in-time inventory control commits management to improving quality control. The greatly improved electronic information concerning the production process made possible by the installation of production communication networks aids in the implementation of statistical quality control. The move towards automation with higher quality control, furthermore, greatly reduces waste, thus increasing the amount of output produced from the input.

The speed at which automation advances depends on the amount of economic competition. Currently Japan has gained the lead in robots, quality control and labor-management relations, while the US leads in software development. The US is keenly aware that not catching up with the Japanese may mean losing those industries which are becoming automated. As a consequence a competitive mentality focusing on a world automation race is developing much like the mentality which powers the arms race. The Europeans are also aware that to be competitive they must follow suit. While the intense competition propels automation forward, progress can not proceed faster than the accumulation of knowledge of production processes. Automating a task requires a task behavior model, which varies considerably in the required depth of understanding depending on the task. Building a control device for a chemical production process, for instances, requires a mathematical model describing the behavior in continuous time. In contrast, building a model of a decision process in order to implement an expert system requires only an understanding of the rules of thumb by which the expert makes the decision.

In short, the important point in transferring human skills to software is that the human knowledge, which is frequently in the form of intuitive skills, must be understood to the level of a formal model which can be made operational in software. If a process is to be controlled in continuous time, the model must be able to simulate the task behavior accurately enough to be controlled by the computer model. If the decision is a one-time decision which is not time dependent, any model which produces better results than all but experienced experts can produce is useful. Success in automating a task, in other words, depends on the level of knowledge needed in order to carry out the task. Because these knowledge requirements increase with the complexity of the task, automation will proceed from the routine to the nonroutine.

The economic costs of developing and using a model to program a task are very different from the cost of training humans to perform the task. Generally it very much more expensive to develop and use a model to program a task, than to train a human to perform the task. But, as the cost of reproducing a program is negligible, the greater the number of people which must be trained, the greater the economic incentives for model and software development. Currently research in robotics and machine intelligence takes places primarily in universities, but much of the major automation innovation takes place under an improvisatory strategy when plants are modernized and when new plants are constructed. Such innovation can be extraordinarily expensive. General Motors is currently spending tens of billions of dollars trying to leapfrog its Japanese rivals by making a major innovation in automation[18]. Frustrated by the rigid work rules of the auto union, GM attempted to bypass labor by using MAP to create highly robotized, paperless factories. The plan was to modernize several plants and then use the acquired knowledge in creating the all new factory of the future in the Saturn plant. GM pushed too far beyond the state of the art in making this attempt and is currently engaged in much applied discovery in how to make the installed automation work up to design specifications. The plans for the Saturn plant had to be scaled back.

Too make matters worse the Japanese management of the joint GM-Toyoto plant in California instituted Japanese labor-management relations and were able to achieve near Japanese levels of production and quality without a major investment in automation. It is far too soon, however, to consider GM's attempt at automation a failure because with successive revisions they are likely to achieve significant results which will make them a better competitor. With Japanese automobile plants in the US, the Japanese firms will probably learn the best from the US software intensive approach to manufacturing whereas the US firms will learn how to adopt Japanese labor-management relations and leaner, more flexible organizations. The long range result of GM's attempt will be a significant advance toward automation with the paperless factory where all equipment in the office and on the factory floor is linked in a MAP network.

Blue-collar work in manufacturing will decline gradually and not precipitously for several reasons. First, the cost of discovery in applying new ideas is so expensive that only a small range of new ideas are likely to be implemented at one time. Imitators can adopt the most successful aspects of the most advanced factories at a fraction of the cost of trying to push the automation frontier forward. Second, the diffusion of automation techniques from the leading industries, will take time because of the enormous range of manufacturing processes and techniques currently employed. To adapt these techniques will take expensive applied discovery and numerous equipment modifications. For example, to adapt a particular type of FMS cell to make a new type of part may require the incorporation of new types of tools and tool motions.

The decline in blue-collar work in existing manufacturing industries is likely to proceed step by step as each new factory or factory modernization displaces an obsolete plant. The construction of new plants in new industries is likely to be at or near the automation frontier. A current growth industry, which is in its initial phase, is biotechnology which is likely to be highly automated while it expands. For example, the production of products using DNA altered bacteria is a continuous process for which current knowledge is sufficient to automate. As the growth of new manufacturing industries is likely to displace older obsolete industries, the growth of total manufacturing output is likely to increase with a decline in blue-collar work.

The advance of automation in manufacturing will change the nature of blue-collar work in manufacturing. As the programming of tasks will proceed from the routine to the nonroutine, work will increasingly become a sequence of unique tasks such as repairing breakdowns and installing modifications to equipment to improve productivity or make new products. Advances in automation also tend to shift work from the actual physical manipulation of objects to the manipulation of representations of the objects in the social nervous system. For example, the automation of the production of parts first with the numerically controlled machine and later with the integration of flexible manufacturing cells with computer assisted design shifts work from a machinist working with objects to a technical person working with programs.

The advance in automation coupled with a much greater emphasis on quality control will gradually increase the educational requirements necessary to perform tasks in manufacturing. The shift of work to the social nervous system means that workers will need some computer literacy to operate equipment and some knowledge of programming to make improvements. The application of statistical quality control requires some knowledge of mathematics. The manufacturing workforce will gradually become technicians. At the same time much work will become highly technical such as the creation of expert systems to diagnose failures in complex equipment such as robots. As knowledge of programing tasks in manipulating physical objects in manufacturing grows, this knowledge will be applied to the programing tasks in manipulating physical objects in industries outside manufacturing such as construction and service industries. For example, advances in robotics have lead to robots which deliver meals in hospitals. Robots have also been developed to clean floors[19]. Consider, for another example, the job of a repairman, whether he works on a defective factory robot or a defective automobile. For diagnosing faults in equipment, expert systems have already demonstrated excellent performance and the task of diagnosing equipment failures will be increasingly programed. For a final example, the preparation of fast food such as McDonald's BigMac will be increasingly automated[20].

The decrease in total hours of employment per week in such blue-collar work outside manufacturing will decrease more slowly than the corresponding measure for manufacturing for three reasons. First, blue-collar work is less routine outside manufacturing than within manufacturing. For example, conditions at a construction site are much more variable than in a manufacturing plant. Significant automation in construction may require a whole new construction technology with modules constructed under controlled conditions in factories and assembled at the site. Similarly, transportation on roadways has many nonroutine elements. Also, the construction of a robot to clean floors must have a much higher level of machine intelligence than a robot to paint automobiles because the cleaning robot must deal with so many more contingencies. These nonroutine elements make the construction of models of the respective processes difficult.

Second, there is much less international competition in construction and services, hence there is less pressure to automate. Third, the service component of the economy is likely to grow faster than the overall economy. One example of a growth industry is leisure. Assuming that the real incomes for most groups in society continue a gradual increase, leisure time will increase as most individuals with increasing incomes will want to substitute some of the increase for leisure. While leisure support services will experience some increase in software support, they will not be nearly as automated as manufacturing, since leisure has fewer routine tasks to program. Another growth industry will be geriatric services for the aging baby boomers.

As automation of blue-collar work outside manufacturing progresses, work will become more technical as the control of physical objects shifts to software in the social nervous system. For example, a technical component of repairwork becomes the construction of expert systems for diagnosing faults. Also, the maintenance and repair of robots which clean floors requires much more skill and education than cleaning floors.

Automation: Information Manipulation

The trend in white-collar work, directly depends on the advance of the social nervous system. As activities shift to the social nervous system, the numerous opportunities for programing tasks will cause a significant decline in the total hours of employment per week in such activities. On the other hand, new tasks and industries will tend to offset the rapid decline.

First, let us consider the consequence of shifting the office from paper to microbinics. What prevents programing many routine information manipulation tasks is, not a lack of knowledge, but the fact that the current information manipulation technology is a mix of paper and microbinics. This mix of microbinics and paper has actually increased the amount of paper and has created more routine white-collar work such as filing and mail service. A significant decline in routine office work will not occur until the current efforts to create standards for representing various kinds of information objects used in office work succeed and are generally accepted. General acceptance of standards will enable the office to shift entirely to microbinics.

Once the office has shifted to microbinics, current knowledge can be applied to rapidly program routine administrative functions such as filing and electronic mailing. Moreover, much more than simple filing and communicating tasks can be programmed as the office shifts entirely to the social nervous system. The current mix of paper and microbinics creates considerable work in order for humans to transfer information objects back and forth from paper to microbinics in the flow of information. In the future such tasks will be programed. For example, the flow of required data and reports between firms and the government will take place in the form of one computer talking to another computer.

When some activities are transferred to the social nervous system, most tasks will be programmed immediately. Consider the shift of a library to microbinics. Once a user has found the item he wants in an electronic catalog, software will automatically transfer the item for viewing. The task of restocking the shelves has been eliminated and the task of checking out material has been programmed. As software to search for material becomes more sophisticated and easier to use, the task of a reference librarian will have been programmed. The tasks that remain are the unique tasks such as deciding what new material to acquire.

Given that the shift from paper to microbinics will not take place in all offices and households simultaneously, dual microbinic and paper systems will be maintained for some time. The complete shift will start with large corporations and government and will proceed to smaller firms and households. In twenty to thirty years, as most households switch to the basic wire ISDN standard and acquire capable home computers the use of paper will precipitously decline.

The impact of programing routine tasks in office administration will be much greater than the impact of gradual automation in manufacturing. Because routine paper handling tasks throughout the political economy are similar, little adaptation is required to develop software for the entire political economy. Thus the substitution of programs for humans in routine information manipulation tasks will rapidly diffuse through the entire political economy.

An additional aspect of programing routine tasks of information manipulation is that the residual labor is likely to be transferred to managers and professionals. With the development of voice-activated natural languages, most managers, who will be trained in making decisions based on data analysis, will operate their own workstations in wordprocessing and decision analysis tasks such as the creation of a nonroutine report. Consequently, as administrative-support software advances, it will be more efficient for the professional to work directly with the software than through a low-paid intermediary and thus the traditional secretary will become a status symbol for the truly elite.

In 1988 approximately 15.7% of all work was white-collar work under the category of administrative support and clerical[21]. In the next forty years this is the area of work upon which automation is likely to have the greatest impact. This type of white- collar work is likely to decline absolutely and the drop is likely to occur rapidly within a period of one or two decades. The positive aspect of this type of automation will be a great increase in office productivity.

As activities are transferred to the social nervous system, programs will displace humans from service tasks by transferring the residual labor to the customer. The oldest example of this trend has been the development of the telephone service, where automatic dialing greatly reduced the need for telephone operators long before the advent of microbinics. A more recent example is the automatic teller machine, ATM. The displacement of human tellers by ATM's would take place much more quickly if banks charged the customer the true price of a human teller transaction. The customer incurs no service charge for a human teller transaction, whereas the customer is charged for each ATM transaction. This inhibits the volume of ATM transactions from obtaining economies of scale.

As retail shopping shifts to the social nervous system, routine tasks will be programmed and the residual labor transferred to the customer. For retail shopping to shift to the social nervous system the customer needs software to compare his alternatives, buy the desired items, and make payment. Also, the customer must be able to physically obtain the purchased items conveniently. As the capacity of the social nervous system for image communication expands, software will provide customers with as much visual information as currently available in live displays and at the same time will enable the customers to make analytic comparisons. With the development of convenient pickup points, the consumer becomes a much more efficient comparison shopper. With such software most of the labor in retail shopping has been transferred to the customer.

As retail shopping shifts to the social nervous system the labor in retail stores is reorganized. As the display of merchandise becomes a slide image or a video, stores will eliminate the task of stocking physical display areas and will become automatic warehouses. The task of operating the cash register has been programmed with the automatic payment system. A few customer representatives would remain to handle only unusual cases through communication channels. Accordingly, the white-collar work week in retail trade will gradually decline. Automatic reordering and automatic warehouses will also cause a gradual decline in employment per week in wholesale services.

The development of a single customer representative, supported by software, who services all of the customer's needs, is an intermediate step in transferring the residual labor to the customer. A good example of this automation is the airline customer agent, using the airline reservation system, to provide the customer with the desired flights. As the public becomes computer literate, most members of society will be able to operate programs using interfaces such as menus or icons. The trend will be to have the customer directly interface with the various programs which are currently controlled by the service representative. While customers can today interact directly with airline reservation systems using personal computers through the information utilities, the programs would have to be simplified for general use.

As software becomes increasingly sophisticated, it can dispense an increasing number of services. One example to consider is the realtor. Currently the realtor in most cities has listings available in a computer memory. To obtain a listing of all properties with a given set of characteristics, all he need to do is to run a simple program on his computer. A loan comparison option is also being added to this type of software. But if the potential customer could directly address the program, which could be made interactive by means of help instructions for even a computer neophyte, the realtor would become superfluous. In a similar fashion many standard tasks of the legal profession, such as wills, divorce, and incorporation could be performed by programs.

The basic reason that this will not happen in the near future, however, is that most professions have some monopoly power over the sale of their service. They will attempt to extract monopoly rent from the use of the programs for their profession as long as possible. In competitive industries, on the other hand, service software is likely to be developed more quickly. Industries such as banks and airlines, where the service representatives are salaried, would have much more incentive to transfer the labor cost to the consumer in order to remain competitive.

The category of work which will undergo the second most rapid decline is white-collar work in sales occupations, which current comprise some 11.8% of all civilian work[22]. This decline will occur after the decline in administrative support, including clerical because a precipitous decline in sales occupations will not occur until households have higher capacity telephone service. As the basic wire ISDN service and fiber optics start moving into household in significant numbers in twenty years, the sharp decline of white-collar work in sales occupations will begin.

As activities are transferred to the social nervous system, most routine tasks will be programed as part of the transfer. For example, the automation of the flow of data within a corporation provides a framework for automating the creation and communication of routine reports. Thus, within such a framework division reports to corporate headquarters and reports to government will tend to be programed and communicated automatically. Transferring marketing data to the social nervous system provides a framework for programs to automatically merge, clean, and maintain customer lists. Orders can be entered electronically such that programs can automatically determine available inventory and subsequently monitor the status of the customer's order[23]. In the field of medicine, online databases such as MEDLINE provide doctors access to the latest research to deal with a rarely encountered disease. The transfer of medical records to microbinics provides a hospital with much more than a more efficient means of bookkeeping. Software monitoring each patient can alert staff members to specified alert conditions such as dangerous side effects to administered drugs.[24].

However, for the transfer of an activity to the social nervous system to be an innovation, the institution usually must redesign the administrative process to take advantage of the new combination of hardware, software and humans. The pace of such transfers depends on advances in hardware, software, communications and a deeper understanding of administrative processes. Nevertheless, once an activity has been transferred to the social nervous system, the programing of tasks proceeds from the routine to the nonroutine. The rate at which more complex, nonroutine tasks are programmed depends on the level of understanding of the activity. This raises the question of how fast white-collar work will decline in such complex tasks as management, producer and other professional services.

The rate at which humans will be displaced from performing complex tasks, such as decision making and professional judgments, depends on the advances made in expert or knowledge-based systems. In decision making, expert systems are better at making such routine decision-making as credit authorizations than an inexperienced credit officer. In accounting, expert systems provide useful advice on tax matters. The medical profession has an increasing number of expert systems to diagnose diseases, some of which are better than a general practitioner. Even though such knowledge based systems can hardly be claimed to "think"[25], they provide an economic function. Their development and use are likely to increase.

Yet such expert systems are very unlikely to completely replace managers and professionals in the foreseeable future. While expert systems in such areas as repair act as a primary expert providing the repairman a diagnosis, the more common use of such expert systems is to act as an intelligent assistant. The less the underlying behavior can be reduced to a comprehensive model the more an expert system is likely to be an assistant for a specialized application. For example, an expert system for loan appraisal would become an assistant for an experienced loan officer. Expert programs in medical diagnoses are likely to be used as a decision aids by doctors. Given the fact that human capabilities such as intuition and imagination are not likely to be reduced to mathematical models any time soon, the advance of expert systems is likely to displace assistants more readily than the principal managers or professionals.

One factor which will slow the decrease in the total hours of employment per week in management, producer and professional services is the growth of new tasks. With a rapid rate of technological advance and market change, managers will increasingly face a sequence of new decisions to accommodate themselves to change. The shift of management from intuitive decisions to analytic decisions using giant decision-support databases greatly changes the nature of management decisions since such decision-support databases enable managers to analytically consider a large number of alternatives in great detail. This approach will be an innovation since the limited cognitive capabilities of humans prevent them from carefully considering more than a couple of alternatives.

Supermarket managers, for example, will examine the profitability of every item on display and move to making forecasts of the profitability of alternative displays based on analysis of the effectiveness of various types of displays. The increase in the analysis of numerous alternatives will consider public as well as private issues such as the environment and safety. The advance in decision-support databases and models would promote a large increase in the analysis of both public and private alternatives. In a similar fashion, professions such as accounting will increase their analyses of alternatives by being able to present many more "what if" scenarios using models and expert systems. Doctors using expert systems as decision aids to diagnosis diseases will analyze the symptoms with respect to all diseases in the database not just those with which the doctor has had recent experience.

At the same time that households begin to perform more shopping through the social nervous system, they too will gradually shift from intuitive decision making towards the analysis of alternatives using household-decision-support systems. A consequence of the use of decision-support databases by individuals for all types of decisions both at work and at home will create a great demand for information which, in turn, will initiate a major growth in all kinds of information services. Some current examples are airline reservation systems and stock market financial data.

At present, the growth of information services is inhibited by the low text and data transfer rates of the analog telephone system into small businesses and the home. But with the advent of the much greater capacity of the basic wire ISDN standard, information services will have a relatively inexpensive conduit for the transfer of large volumes of data, text, and slides. And as local fiber optic communications becomes a reality, these data services should become really sophisticated.

The growth of such information services, however, will not create a large increase in white-collar work. Such services will be dispensed as software programs with the residual labor transferred to the customer. Humans will labor in creating better software and additional services to add to the databases.

In addition to the creating of databases to support decision making, the expansion of the social nervous system will create opportunities for the creation of a whole range of image and text services such as libraries of text and videos. The growth of a new interactive media involving all types of representations of information objects such as text, voice, image, video, data and symbols has just begun. The promise of this composite media, sometimes called multimedia or hypermedia, is that the products are capable of interacting with the viewer to provide multiple responses to meet the needs of alternative individuals[26]. When these new information services come into being, they will use the latest software technology. This means that these new services will be at the frontier in the use of programing tasks to expand rapidly without needing a large labor force.

Another growth industry will be adult education to retrain adults displaced from work. With a high rate of technological advance, new goods, services and production technologies will be constantly displacing old goods, service and production technologies. The old activities will experience a rapid rate of job destruction while the new activities will experience a rapid rate of job creation. At the same time soft automation will be constantly reorganizing the workplace. The number of levels of managers in institutions will decrease. Also, work in groups such as professional groups will consist of smaller groups supported by more software. Moreover, as soft automation frequently advances in programing tasks rather than entire jobs, work is transferred to the social nervous system and reorganized in a manner which requires new skills. To maintain continuous employment in such a work environment, adults will have to be constantly learning new skills. This demand for learning will greatly expand adult education.

White-collar work should gradually decline after thirty years. At first shifting activities to the social nervous system will create an immediate decline in work as the routine aspects of information manipulation are programmed. Further advances in soft automation will frequently be slow because of the lack of knowledge to model social processes. As the routine tasks in information manipulation are programmed, white-collar work will require much higher skill levels. For example, work as a secretary will be transformed into the job of a multimedia editor. Routine data manipulation will be performed by adults who are sufficiently computer literate to use many software programs.

Income Distribution

Before considering the problem of forecasting changes in the income distribution, let us consider the simpler problem of maintaining full employment. This depends on more factors than just the rate of job destruction from automation and new job creation from new industries. Changes in the population structure and the number of legal and illegal immigrants affect the demand for jobs as well. A population bulge, such as that produced by the recent baby boom, entering the labor force creates a large increase in the demand for jobs. Foreign competition, too, can displace large numbers of jobs or even entire industries. The supply and demand for jobs is equated not only by the level of financial compensation, but also by the length of the work week.

Government policies can, to a limited extent, help to maintain full employment. With an increasing rate of discovery, invention, and innovation the rate of both job creation and job destruction will be high. Unfortunately, neither the creation of new industries nor the displacement of jobs through automation proceeds on a smooth, predictable path. Most individuals will become unemployed many times during their careers and each time, they will have to find new jobs. Government, by creating retraining programs either alone or in conjunction with the private sector, can reduce the unemployment rate by shortening the transition time for finding a new job.

In addition, there are limits to the extent that government can reduce unemployment through stabilization policy–that is stimulating the economy through monetary and fiscal policy. Such a policy will produce inflation in expanding industries with shortages of skilled labor without reducing the unemployment of individuals lacking desired skills. Rather than try to train the least capable in an advanced society, employers will have incentives to find more capable, more highly motivated labor in less developed countries. For information manipulation tasks, foreign workers can be imported into firms via teleconferencing[27].

The government can also promote full employment by taking steps to reverse the temporary decline in US competitiveness. First, primary and secondary education can be improved to produce at least as qualified a workforce as our competitors. The school year is only 180 days in the US in contrast to 220 days in Germany and 240 days in Japan. Second, the cost of capital in the US has been much higher than our rivals. To reduce the cost of capital Congress will have to eliminate the budget deficit and provide incentives to increase the low savings rate. Third, the US needs to maintain research and development expenditures at a competitive level. Finally, the finance mentality of MBA's needs to be broadened to focus on the problems of production.

Assuming the government has the leadership to address these problems, there remains an important long term trend. With advances in automation the total hours of employment per week to produce an increasing amount of goods and services will gradually decline. With increasing population full employment requires a declining workweek. For example, if the standard work week declined instantaneously from forty to twenty hours, then twice as many workers could be employed in a five day week.

The natural reduction of the workweek depends on the rate of productivity advance. From the perspective of management, the prospect of a decrease in the work week can be very threatening, since workers may demand the same or greater real wage for fewer hours of work. However, if productivity advances faster than the decrease in the work week, then firms can simultaneously decrease the work week and increase income without increasing labor costs. Employees are generally quite happy to receive higher incomes and more leisure. Such was the case from the Civil War to 1940, as the average work week fell from 67 hours to 40 hours. Since 1940, however, the rate of productivity increase has declined and the average work week has declined very little.

Fringe benefits, of course, create negative incentives for managers to reduce the workweek. From the perspective of management, the cost of a fringe benefit such as medical insurance is reduced by having a smaller labor force work longer hours. To circumvent this problem, firms seeking to cut labor costs and achieve greater flexibility in changing market conditions are moving to a dual labor force. A core workforce with skills essential to the competitiveness of the firm is employed full time with fringe benefits. The second group of workers, however, are part-time workers and temporaries who do not qualify for fringe benefits. It is through the increase in such part time employees that the decline in the workweek is taking place.

Automation is likely to make a substantial improvement in productivity, especially when productivity is given a broad definition including such aspects as increased flexibility to react quickly to changing market conditions. For those individuals who constantly upgrade their skills to meet the requirements for jobs in the constantly reorganizing economy, their work should exhibit constantly increasing productivity. Assuming the economy maintains a high level of investment, the forecast is that real incomes will gradually rise concurrent with a declining work week[28].

The final question to consider in the income distribution is whether the income distribution will become more or less even. The advance of automation implies that, over time, the production of most goods and services will become more capital intensive primarily through more software per worker. For the past one hundred years the growth of services were primarily labor intensive. Recently services are becoming capital intensive[29] and in the next 100 years, they will become even more so with much greater use of software. With capital, that is hardware and software, an increasing component of production, the income share accruing to capital will gradually increase. As the ownership of capital is concentrated, the increase in the income share to capital means the income distribution will gradually become more uneven as the rich will acquire a greater share of national income.

A second factor leading to a more uneven income distribution is the nature of international competition. As more third world countries industrialize, international competition will intensify. With advancing automation, work will tend to become a sequence of unique decisions involved in adapting to constantly changing technology. With intensive competition, the reward system will favor the ability to produce rather than cronyism. Those with special skills which cannot easily be automated will obtain rewards equal to their economic value. The income distribution will proceed towards the bell curve of native talent. Family income will become even more uneven as women achieve equality of pay with men, because successful women tend to marry successful men. The less able are more likely to have more routine jobs which are more likely to be reorganized with each advance in automation. Thus the less able will have decreasing incomes in periods when advances in automation temporarily increase the rate of job destruction in relationship to job creation.

The increasing unevenness of the income distribution will create a problem of political stability. The assumed economic condition for political stability is that almost all groups in the political economy receive increasing real income over time. With advancing automation this condition is not likely to be met. As the the income distribution becomes more uneven, the annual real increases in income to the less fortunate will be small or even negative. In periods of rapid reorganization in the political economy, such as will occur when the political economy shifts from paper to microbinics, large numbers of workers will be negatively affected. The less fortunate will resent their deteriorating relative position. With automation, then, the empirical condition for long run political stability is unlikely to be met, especially during periods of rapid reorganization.

In the 20th century the political solution of industrial democracies to achieve political stability has been the creation of the welfare state. In information society there will be a need of a welfare system to ease the problems of adjustment to advancing automation and to provide incomes to the unfortunate. The goal in designing a welfare system for informational society is that it should have desirable properties. First, the welfare system should cover current fringe benefits so that managers will not have incentives to resist a decrease in the work week. A smooth decline in the work week will serve to promote employment.

Second, the welfare system should be not be subject to political pressures for change if over fifty percent of the population might need to receive some assistance. For example, if the welfare recipients become the majority, the politicians vying for political favor should not have an incentive to increase the benefits without limit. As politics operates according to a time frame dictated by the next election, the increased consumption expenditures are likely to be paid for at the expense of private investment. Economic growth will gradually slow down and even might become negative as obsolete plants and equipment are not replaced. This possibility leads to the final condition that the welfare system should provide positive incentives for economic activity.

Trying to ensure increasing real incomes for all through expanded transfer payments fails to satisfy the stipulated conditions. Presently the government is reluctant to provide a single guaranteed income; instead, it provides special services such as food stamps and medicaid. While such a system of specialized benefits could cover current fringe benefits such as medical insurance, such a system creates incentives for increases without limit if over fifty percent of the population receive some benefit.

The solution to the problem will require more than a negative income tax[30]. With this type of income tax there is a boundary income; if a person's income is higher than this boundary income, he pays the government a tax, whereas if his income is lower than the boundary, the government pays him a supplement. The negative income tax is designed such that each individual has a floor on his income. This plan, while a welcome simplification of the basket of special services provided by government, nevertheless suffers the same defect. When more than half the population is receiving the floor income provided by the negative income tax, the political pressures become immense to increase this floor.

Social Inheritance

To place the long-term problem in perspective, assume that all labor is automated out of the production of goods and services. Even if all labor could be automated, bounded rational neoclassical economic principles would still be applicable. The income is simply distributed to capital, that is physical capital and software, and to the owners of raw materials. Market signals still operate as modified by government regulation. Provided the level of investment remains at 20 to 25% of the gross national product, the total output will continue to grow at 2 to 3% a year. To earn an income, then, one would either have to own capital or own a natural resource. Individuals could, of course, increase their capital stock by constantly moving their assets to growth industries, but the political stability of such a system would depend on the initial distribution of assets. If the distribution of assets were not too uneven, such a system would be politically stable.

In considering the transition to a very capital intensive system of production it is desirable to avoid the errors of nineteenth century solutions such as socialism, that is the public ownership of the means of production. The goal is to preserve capitalism with a income distribution that makes everyone better off. One method of accomplishing this objective is to substitute a social inheritance[31] for a private inheritance.

Social inheritance is defined as a system of inheritance where each year the federal government would sell the assets of those who died that year and distribute the proceeds to each state in proportion to the number of citizens. To prevent individuals during the last years of their lives from trying to consume all their accumulated assets, the federal government would install a sharply progressive tax on consumption based on the sale of assets. In addition, as will be discussed in Chapter Nine, each state would administer the social inheritance program for its citizen residents. As each state would be assuming federal welfare programs, each state would have to decide what portion of their social inheritance citizen residents would have to commit to required expenditures such as medical insurance and retirement investments and what portion they could freely spend.

Social inheritance has two attractive properties. First, as the share of income accruing to software grows and the share accruing to labor declines, the income stream from the social inheritance would increase, compensating for the decline in nonspecialized labor income. Second, social inheritance has a very different incentive effective on the political process than the guaranteed income or other income supplementing approach, for the government does not try to shield the individual from the vicissitudes of the market. In general equilibrium theory, without externalities there is a unique competitive equilibrium for each distribution of wealth. Social inheritance as the basis for welfare separates the impact of welfare from price signals.

Whether the substitution of social for private inheritance would adversely affect capital formation is a topic which needs some exploration. In the *Gospel of Wealth*[32], Andrew Carnegie argues that the children of an industrialist should not inherit their parents' wealth because it would dull their incentives to create their own fortune. The industrialist should instead donate his wealth to socially productive activities, for example libraries. Social inheritance makes the donation to society compulsory.

Nevertheless, conservatives would counter that a man should have the right to dispose of his wealth as he desires. To deny a person this right, most conservatives would maintain, would seriously impair investment incentives. If one views entrepreneurship as a game which is intrinsically interesting, however, then the principle motivation is playing the game well, not leaving a fortune for one's offspring. For game players the decrease in investment from lack of an opportunity to dispose of the estate might be quite small.

While social inheritance might not adversely affect the incentives of entrepreneurs, social inheritance would adversely affect the level of investment. In a capitalist society, the level of investment by individuals increases with increasing income. Converting from private to social inheritance would transfer assets to individuals with much lower levels of income. Many of these individuals would want to immediately consume the freely spendable portion of their social inheritance.

To compensate for this decrease, taxes would be shifted from income to consumption[33]. Because almost all transactions will be processed through the social nervous system, the problems with a flat or graduated consumption tax would be no greater than the problems with income taxes today. Such a tax system would eliminate the double taxation of corporations and stock holders. Shifting the tax burden to consumption greatly increases the incentives for investment.

The design proposed in this book is based on the following assumptions concerning the income distribution and workweek. With the implementation of social inheritance, almost all groups in society will enjoy rising real incomes promoting political stability. Nevertheless, even with social inheritance the income distribution will become more uneven. With the advance of automation in both manufacturing and services, the rate of productivity advance will increase to better than 3% in the production of goods and services. Part of the increase will be consumed as increased leisure and part as increased real incomes. In the next one hundred years the average workweek should fall to about twenty hours while real incomes should increase from between two and three times the current levels.

The great increase in leisure creates a new problem. The decreasing work week means people will have to find other activities to occupy their time. For most people in the work force today an important aspect of life is the weekday trek to the workplace. The workplace for most people is much more important in their lives than simply a means of obtaining an income. Besides providing a source of income, the workplace provides a web of social relationships for most people. Moreover, for individuals who regard work as a career, workplace relationships are a source of status and competition. To create a congenial, healthy workforce, many organizations support a variety of sports and social organizations for employees. Thus, as the workweek declines, the role of the workplace in organizing leisure time outside work will decline correspondingly. The design of a political economy based on automation of goods and services, then, must consider explicitly the problem of how people are going to organize their leisure time.

Notes and References 4

1. Marx, Karl, 1867, *Das Kapital: Kritik der politishchen Oekonomie*, (Verlag von Ott Meissner: Hamburg)

2. See the October 1987 and September 1991 issues of *Scientific American* for a forecast of hardware, software and networks.

3. Two of the Bell operating companies plan to have ISDN 90% installed by 1994 and the other five 50% installed by then. See: Bell, T., 1992, Telecommunications, *IEEE Spectrum*, Jan, pp 36-38

4. Timms, S., 1989, Broadband Communications: The Commercial Impact, *IEEE Network Magazine*, July

5. Lubliner, O., 1989, ISDN Development in France, *Telecommunications*, Jul, pp 19-24

6. Seghers, F., 1989, A Search-and-Destroy Mission-Against Paper, *Business Week*, Feb 6, pp 91-95

7. Interestingly enough the legal profession is slowly creating precedents for contracts without paper even in the absence of a secure one-time write device. See: Wright, B., 1992, Contracts Without Paper, *Technology Review*, Jul, pp 57-61

8. Rothfeder, J, 1989, Neither Rain, nor Sleet, nor Computer Glitches.., *Business Week*, May 8, pp 135-139

9. Mitchell, R., 1985, Detroit Tries to Level a Mountain of Paperwork, *Business Week*, Aug 26, pp 94-96 The growth of EDI to drastically reduce paperwork costs in such applications as ordering parts depends on the advance of the social nervous system. See: Cerf, V., 1991, Prospects for Electronic data

interchange: the full value of EDI will be realized when certain information infrastructure are in place, *Telecommunications*, Jan, pp 57-61

10. Woodworth, J., 1988, 1988: The Year of the Debit Card?, *The Bankers Magazine*, Jul/Aug, pp 34-38. In time real-time microbinic money will replace the current payment mechanisms for most purposes. For a discussion of the electronic money revolution see: Solomon, E. H.(ed), 1991, *Electronic Money Flows: The Molding of a New Financial Order*, (Kluwer Academic Publishers: Boston)

11. Some examples of current advances towards the complete microbinic library, see: Motley, S. A. 1989, Optical Disc Technology and Libraries: A Review of the 1988 Literature, *CD-ROM Librarian*, May, pp 8-24 and Hearty J. A. and V. K. Rohrbaugh, 1989, Current State of Full Text Primary Information Online with Recommendations for the Future, *Online Review*, Vol 13 No. 2, pp 135-140. Carnegie-Mellon University is in the process of creating an electronic library network through which students and faculty can obtain articles at their computer terminals. In the future they will be able to obtain books. See: Alexander, M., 1992, University library enters in formation age. *Computerworld*, Mar 2, pp 31. As libraries shift to the social nervous system books will become hypertext. See: Reynolds, L. and S. Derose, 1992, Electronic Books, *Byte*, Jun, pp 263-268.

12. Schwartz, E., 1991, Adventures in the on-line universe. (Compuserve, GEnie and Prodigy), *Business Week*, Jun 17, pp 112-113. Short term public acceptance of these services has been dampened by allegations of censorship and invasion of privacy by the Prodigy management. See: Staff, 1991, Big brother or big brother? (Users accused Prodigy, an information network linking computers of invading their privacy,*Time*, May 13, pp47 and Lewis, P., 1990, On electronic bulletin boards, what rights are at stake? (Prodigy's policy of screening electronic mail seen as censorship), *New York Times*, Dec 23. Interactive video communications will probably to necessary before such services saturate the potential market of every household.

13. See reference 4.

14. The phone companies are ready to deliver such services once the restrictions of the 1984 Cable Act are lifted. See: Cole, A., 1992, Telcos poised for video delivery, *TV Technology*, May. Allowing the phone companies to deliver video services to homes would vastly accelerate the shift from a copper wire to an optical fiber local communication system. This would would have externalities. The downside is that such a move would create a conflict of interests for the phone companies as common carriers providing service for all and sellers of services with incentives to stifle the competition. In July of 1992 the FCC allowed the phone companies to use visual signals.

15. For example, see the special report, Multimedia, in the Mar 31 issue of *PC Magazine*

16. Research in teleconferencing computer terminals is discussed in Brittan, D., 1992, Being there: The Promise of Multimedia Communications, *Technology Review*, May/Jun pp 42-50.

17. Bureau of Labor Statistics, 1988, Employment and Earnings, Vol 35 No 6. Jun

18. The saga of GM's attempt to endrun its Japanese rivals has received extensive media coverage. For example, see: Poe, R., 1988, American Automobile Makers Bet on CIM to Defend Against Japanese Inroads, *Datamation*, Mar 1, pp 43-51 and Hampton, W.J. and J.R. Norman, 1987, General Motors: What Went Wrong, *Business Week*, Mar 10, pp 102-110 Since realizing that the investment in technology would, at best, reap long term benefits, GM has been struggling to create a much leaner, dynamic organization with a strong emphasis on reducing costs to world standards. See: Treece, J., 1992, The board revolt: Business as usual won't cut it anymore at a humbled GM, *Business Week*, Apr 20, pp 30-36; Shiller, Z., 1992, GM tightens the screws: Only the fittest of its suppliers will survive,*Business Week*, Jun 22, pp30-31: and Moskak, B., 1992, GM's new-found religion, *Industry Week*, May 18, pp 46-52

19. Bylinsky, G., 1987, Invasion of the Service Robots, *Fortune*, Sep 14, pp 81-88; Myers, Frederick S.,1990, Play it again, WABOT; Japan's robots aspire to service-sector jobs, *Scientific American*,May, pp 84-85; and Stix, G., 1992, No tipping, Please, *Scientific American*, Jan, pp 141

20. For an example of the effort to automate the preparation of fast food, see: Roboburger, 1988, *Discover*, p 6

21. See reference 16.

22. See reference 16.

23. Moriarty, R. T. and G. S. Swartz, 1989, Automation to Boost Sales and Marketing, *Harvard Business Review*, JanFeb, pp 100-108

24. Rennels, G. D. and K. W. Nevew, 1987, Advanced Computing for Medicine, *Scientific American*, Oct, pp 154-161

25. Dreyfus, H. & S., 1986, Why computers may never think like people, *Technology Review*, Jan, pp 43-61

26. For example, see the state of the art section of the December 1991 issue of *Byte* on multimedia.

27. Importing information workers via teleconferencing may become one of the major trade issues of the next century. See: Pelton, J. N., 1989, Telepower, *The Futurist*, Sep-Oct, pp 9-14

28. Over the past 20 years the workweek for many workers has increased. For example, see: Schor, J., 1991, Workers of the World, Unwind, *Technology Review*, Nov/Dec, pp 25-32. In this book it is assumed that the failure of the workweek to decline is a temporary phenomenon due to the low increase in productivity in the US during this period and increasing world competition.

29. Quinn, J.B., J. Baruch and P. Paquette, 1987, Technology in Services, *Scientific American*, Dec, pp 50-58

30. Friedman, M, 1962, *Capitalism and Freedom*, (University of Chicago Press: Chicago). His ideas have been subjected to an experimental test. For a summary see: Stafford, F. P., 1985, Income-Maintenance Policy and Work Effort: Leaning from Experiments and Labor Market Studies in Hausman, J. A. and D. A. Wise (ed) *Social Experimentation*, (The University of Chicago

Press: Chicago)

31. The need for a device such as social inheritance may not materialize for some time because forecasting the income distribution is a perilous task. For example, the media and most liberal economists believe that the poor have become worse off under the Reagan and Bush administrations. However, my colleague, Dan Slesnick, has shown this is not so. See: Slesnick, D.,(forthcoming), Gaining Ground: Poverty in the Postwar United States, *Journal of Political Economy*. Because the concept of social inheritance is peripheral to the main themes of the book, it could be deleted with few changes.

32. Carnegie, Andrew,1962, *The Gospel of Wealth*, (The Belknap Press of Harvard University: Cambridge). Of course, Carnegie argues that the successful should devote themselves to philanthropy after acquiring great fortunes. The approach taken in this book is that value of the created empire should pass to social inheritance

33. The advocacy of shifting taxes to consumption has a long history. For a recent example, see: Mieszkowski, P., 1980, The Admissibility and Feasibility of an Expenditure Tax System in Aaron, H.J. and M. J. Boskin (ed), *The Economics of Taxation*, (The Brookings Institution: Washington)

Chapter 5

The Community

Introduction

Currently the standard work week is 40 hours. With a commute time of 0.3 hours the average worker spends about 38% of his waking time involved in work. As automation advances, we have assumed the workweek will gradually shorten, and for the purpose of discussion of informational society, we have specified the average workweek would be 20 hours. As the workweek declines, most workers would want to organize their work-time to obtain longer periods of uninterrupted leisure. For example, if a worker concentrated his forecasted 20-hour workweek into two ten-hour shifts on consecutive days, such a worker would spend only about 19% of his waking time in work and would have a five day weekend each week. As the workweek is reduced, leisure will replace work as the primary focus of life for most people. The decline in the workweek raises the question of whether the supply of leisure activities will grow rapidly enough to match the increasing demand.

Today, an individual living in a metropolitan area has wide variety of leisure activities from which to choose[1]. He can visit with friends or relatives in his home, for instance, and additional social activities take place at pubs, clubs and churches. He can read a newspaper or book, listen to stereo, or watch television. Participatory sports are available in clubs and at parks where people can also jog, picnic, enjoy water events, or camp. Even driving and walking are leisure activities for some people. To develop skills a person can get involved with crafts such as pottery or gardening. There are professional sports, theater, museums, and events such as rock concerts from which to select. Although the range of possible leisure activities increases with income, most people of all income levels have a varied menu of possible leisure activities from which to choose.

To forecast whether the supply of leisure activities will expand to meet increasing demand we must first consider the historical record. From 1860 to 1952 the average work week fell from 67 hours to 42 thus increasing the average worker's time awake for leisure by about 22% . Many of the contemporary leisure activities listed above many did not exist in 1860, but with increasing real incomes and advancing technology newspapers, books, radio, stereo, and television became available to most people. The number of professional sports has also continued to grow and to spread to all major metropolitan centers, since the 19th century creation of professional baseball. Parks and other recreational facilities have been built since the turn of the century.

Moreover, as technology advances, new leisure applications are invented. For example, computer games are a recent arrival on the recreational scene, and while these games are currently in a decline as a fad, the longterm prospect of using artificial intelligence to make simulation games which are intrinsically interesting will undoubtedly renew the demand. These new activities to some extent displace older activities, much as computer games have now largely displaced pin ball machines.

Most institutions in society had some role in the creation of more leisure activities[2]. Some of these activities were created by individuals for personal use. Others were created by nonprofit associations such as the Boy Scouts and still others, such as newspapers, radio, and television were created by entrepreneurs to sell as products. And finally, governments at all levels have created parks and recreational facilities. In the past the need to create more leisure activities, for instance, public parks, while a political concern has never been considered a national emergency.

Similarly, in the future, provided the work week does not decline too quickly, the current institutional structure will probably generate sufficient activities to occupy the additional time released from work. For example, one growth component of leisure will be adult education selected by workers to prepare for the next job. It is assumed that sufficient incentives exist to create leisure activities without any major new governmental initiatives. As the work week declines, the number and variety of nonwork activities is likely to continue to grow.

Currently a lifestyle in which leisure rather than work is the primary focus exists in successful retirement villages and among certain primitive tribes[3]. The issue, therefore, is not so much whether it is possible to design informational society in which leisure activities predominate, but how to achieve a desirable design. Such a design must, of course, meet a variety of criteria. One important factor in organizing leisure activities is location. To ensure that the location of leisure activities is desirable to the participants, individual choices should play a major role in determining where leisure activities are located. A second criterion is to approximately minimize the transportation costs of individuals in transit between leisure activities. To the extent that individuals bear the transportation costs, they will have an incentive to reduce those costs. A third criterion is the need to organize leisure activities in such a way as to promote a sense of community, for if the importance of the workplace decreases in integrating individuals into society, then some alternative institution is needed to perform this function. As leisure time becomes much greater than work time, a logical choice is the institutional structure for providing leisure activities. If the resulting design succeeds in reducing loneliness, it should be desirable to individuals.

Current use of transportation

A starting point for considering the institutional design for organizing leisure activities is a consideration of how the current transportation system should be employed in leisure activities. The physical distribution of economic and leisure activities in modern metropolitan regions has been made possible primarily by the development of the automobile and the system of roadways. With low-cost energy, the automobile has offered the public a device which could traverse any roadway at the convenience of the owner.

The creation of a roadway system enabled individuals to rapidly traverse between most points in the metropolitan area, giving individuals much greater freedom of choice in deciding where to reside in relationship to the place of work. In the 1920s this freedom of choice led to the creation of suburbs surrounding the major cities[4]. And after World War II, a coalition of interests built the urban freeways which greatly expanded the suburbs surrounding the major cities. With the automobile, suburbia became the dominant type of community organization around the cities. At the same time, the automobile, truck and communication advances decentralized the central business district into numerous satellite centers[5].

Because the roadway system is designed to cope with the morning and evening rush hour traffic, the system in the off hours is efficient in providing people access to their chosen leisure activities. For the consumer this system minimizes his transit time, for instance, to an arbitrarily chosen shopping center. In suburbia, then, leisure activities can be arbitrarily located, as they can be quickly reached by automobile in nonrush hour traffic.

The organization of society based on the automobile provides the individual with more choices than previous social organizations could provide. An individual can opt to live near the city center in an apartment or condominium or further out, he can select a city residential neighborhood. Even further out are the options of suburbia, and for those willing to drive a considerable distance, a quasi-rural location is possible. In addition, the shift of offices to satellite centers on the metropolitan beltways has increased the attractiveness of quasi-rural locations. An individual's choice is thus limited only by the availability of residences in his price range and by the time he is willing to commute. In selecting his residence, an individual need only consider location dependent attributes such as the crime rate, the quality of the public schools, and the prestige of the neighborhood.

Many aspects of an individual's leisure activities are not dependent on his or her neighborhood, because even if they are located a considerable distance from the neighborhood, they can readily be reached by automobile or obtained through the communication system. An individual, for instance, can drive a considerable distance to a shopping mall, sports center, or church. This means that an individual has a much greater choice in his lifestyle than the options offered by his immediate neighborhood. With the advance of television and satellite communications, all households, regardless of how isolated, can tune in to the mass culture. For individuals whose economic relocations are frequent, then, the current social structure should be considered a significant advance over the pre-automobile society. To enjoy his lifestyle the frequent mover only has to find a residence in the general area of his desired activities and not in a specific neighborhood. His market choices in obtaining a residence are accordingly greatly expanded.

While the automobile has provided the individual a lifestyle replete with choice, this form of social organization has many direct and indirect costs. First, automobiles consume large amounts of energy and pollute the atmosphere[6]. Yet another problem created by automobile which is harder to evaluate numerically is the problem of strangers[7]; a lifestyle organized around the automobile inhibits interpersonal relationships in neighborhoods by reducing natural meetingplaces, and with the increased traffic volume linking the suburbs to the central business districts, the quality of life in older neighborhoods traversed by major arteries has been reduced. The fact that the automobile provides easy access to all parts of the metropolitan regions has, furthermore, increased opportunities for theft because the neighborhood has been desensitized to the presence of strangers. In addition, the evolution of most metropolitan regions has resulted in a central city surrounded by contiguous satellite cities and towns of various sizes. With few exceptions there is no metropolitan government to coordinate those services, such as water and sewage, which are most efficiently provided by the entire area. At the same time the various existing local political entities, such as towns and cities, are generally much too large to provide much sense of community.

Any proposed improvements to the contemporary social organization must consider the future demand for local travel. First, consider the demand for travel for business purposes as the automation frontier and the social nervous system advance. As the average workweek declines the peakload demand for a massive system of roadways to move people on a daily basis to and from work will decline. Suppose the workweek falls to two days. If the economic institutions are operated four days a week with two shifts of two days each, the number of people demanding transportation on any given day is halved. For those who must travel to a fixed location, however, the average distance between the workplace and residence will probably increase with declining workweek. For example, with a two-day workweek, an individual may choose to stay overnight near his workplace and commute a considerable distance. Such a choice would allow the individual to reside near his leisure activities, which occupy 5/7's of his week.

A second factor in the decreasing demand for business travel is the increasing number of jobs that will involve the processing of information. Currently many businesses are innovating new teleconferencing and telecommuting applications. As such capabilities of the social nervous system expand, communication will increasingly substitute for travel in business. With inexpensive dynamic image communication, for example, increasing numbers of group meetings will take place through image communications. Many information workers will become fully functional at home or at a local office, thus greatly reducing their demand for work-related transportation facilities[8]. The demand for travel for business purposes, therefore, should steadily decline.

The demand for travel for leisure will also be influenced by the improvement in communications. Improving communications will cause substitution of communication for travel in leisure as well as business. As will be explained more fully in the next chapter, many market activities will gradually move to the social nervous system which means many purchases will take place at home, eliminating much of the driving that is currently undertaken for shopping. As image communications decrease in price, group meetings by communications will become commonplace for leisure purposes. As leisure and wealth increase there may well be an increase in pleasure driving; however, this should not become a major factor in the demand for travel. The bulk of leisure transportation will derive from travel in order to participate in activities where physical presence is required. Such activities in which people physically participate are generally small-group activities, such as team sports, which rarely involve more than thirty participants at one time. Arts and crafts are small group activities. The duration of leisure activities is generally much shorter than the length of the workday. If many individuals participated in activities widely scattered throughout the metropolitan area, the demand for travel could increase with increasing leisure.

The current automobile transportation system could conceivably be used to facilitate the small-group meetings of informational society. In participating in various small-group activities people could traverse to various parts of the metropolitan area. This approach, however, would involve significant energy and pollution costs and would not deal with the problem of strangers. To reduce the transportation costs, it would be efficient to organize society in such a way as to locate in the areas where people live most of the leisure activities that require people to interact directly with each other or with physical objects.

System of local government

The first step in promoting this local organization of leisure activities is creating a system of local government. Since most people live in or near major metropolitan centers, let us focus our attention on modifications to the existing metropolitan political institutions. The function of the metropolitan government is to coordinate the activities of the metropolitan towns and to provide those services for which there exist metropolitan economies of scale. To better accomplish these purposes, the governance of metropolitan concerns should be consolidated into a single metropolitan government, and the governance of local concerns should be partitioned into much smaller towns. Such an organization would create political units more responsive to local needs than the current patchwork of cities and towns. The towns which are subordinate to the metropolitan government, govern local issues such as neighborhood zoning, leisure activities, and the local economy. The new towns, which would rarely have more than ten thousand inhabitants, are conceived as a possible evolution of residential community and neighborhood associations[9]. As has been demonstrated by the examples of Toronto and Minneapolis/St Paul, which have already established metropolitan governments, the creation of metropolitan governments is politically feasible. The partitioning of the metropolitan area into small towns, however, would be more difficult.

The main difference between the design and the current metropolitan organization is that the metropolitan government would directly control activities on land that are not controlled by local town governments. The sites of large-group activities, such as the central business district, major shopping centers, factories and metropolitan parks, would tend to locate on metropolitan land to avoid the vicissitudes of local town zoning restrictions. Part of the land under direct control of the metropolitan government would be green zones between each town. The purpose of such zones would be to provide each town with a distinct physical identity and to provide for major roadways and other conduits such as oil pipelines. For such land use, a local permit from only one government would be required to transit a major metropolitan area. The policy of metropolitan land use would tend to be market oriented, whereas, in contrast, town land use would be oriented towards maintaining the quality of life in the neighborhoods by providing local leisure activities.

As the number of leisure activities expands, of course, the individual must make choices, since, for example, he can not live more than one lifestyle at the same time. Likewise, he lacks the time and resources to meaningfully pursue all the possible activities. Take, for instance, sports. Meaningful participation in a sport requires an investment of time to learn the rules and to master the skills required. In order to be competitive most individuals are limited to participating in a small number of the recognized sports. The same argument applies to arts and crafts or intellectual activities. While people could vary their activities on occasion and could access a wide variety easily, for this design to succeed, most people would spend the major portion of their time involved in the activities physically available in their town and through the social nervous system. The design conditions would be met if individuals lived in towns which supported their lifestyles.

As in the case of each individual, each town would also face a problem of choice, since a town can not simultaneously support several incompatible lifestyles[10]. A town would be limited in space and resources in the number of activities which could be supported. As many activities require sizable investments in space and resources to provide the facility-for example, golf, swimming or automechanics-the members of a town numbering in the several thousands would be able to offer only a fraction of the facilities for all possible human activities. This does not mean that an individual would not have access to a much wider range of activities than physically occurred in his town. Through the social nervous system he would have access to all the activities which took place in this medium. Moreover, a person willing to incur some transportation cost would have access to all the activities which took place in the metropolitan region.

To promote the lifestyle of the majority, the town is to decide what activities can take place in public places. Given the wide range of possible human activities not all activities can take place in a finite space in a finite time. The town must therefore choose what activities it wishes to promote. One choice mechanism is the market; people could use the land not zoned for residents in any manner they desired within the town guidelines for externalities. As automation gradually proceeds, leisure activities would increase relative to market activities. This means that more and more concern will focus on matters of lifestyle or nonmarket criteria. At the other extreme, a town might require all individuals and groups with proposals for nonresidental land use to obtain a specific charter from the town government. The town government would then determine whether the proposed land use fitted the specified conception of the town lifestyle. The point is that the town would have the right to decide what physical activities taking place within its boundaries were in the town's interests. A town promoting a gay lifestyle might have a strip set aside for gay bars, whereas a town supporting a fundamentalist Christian lifestyle might specifically prohibit bars, massage parlors, and porno shops.

To further promote the lifestyle of the majority, towns would have broad powers over the physical organization of the town. Currently in most major cities local neighborhood associations battle outside economic interests for control over the character of the neighborhood. With the partition between town and metropolis, most large institutions would locate on metropolitan land thus easing much of this friction. Within the town the majority would have control over both the types and the location of buildings. The majority would also determine the local transportation system. The town would decide both the location and the amount of physical space to be allocated to sites for various activities, such as residences, the town market and leisure facilities. If a town should desire only single-family homes, the town could so decide. As it is well known that alternative arrangements of human habitats affects the quality of human interactions, the town in making such decisions is determining how much and in what way the physical organization facilitates the human interaction[11]. The difference between current practice and the proposed design is that since the town would be a small political unit, such decisions would much more reflect local and not outside economic interests.

In current society, relocations are primarily determined by economic opportunity. As people gain greater freedom of location through shorter workweeks and as a greater percent of activities take place through the social nervous system, this reason for moving will decline. Increasingly, the reason for moving will be to find a town which supports a desired lifestyle and a neighborhood where an individual is welcome. Given the assertion that people will move to find a town which supports a particular lifestyle, what types of towns are likely to result? First let us consider the general nature of changes from current arrangements. One of the main reasons that families left the central cities for suburbia after World War II was to find a single-family home with more space, privacy and security for raising children. In these relocations, considerations outside those directly related to the immediate home were secondary[12].

In informational society, too, the desire for more space and privacy will be important. With the advance of automation in housing construction the goal of owning a single family home will become a realizable goal of most families. As most people are increasingly freed from a specific location metropolitan areas will tend to spread out, and the choice mechanism will operate to emphasize those activities in which a town excels. If a town excels in its sports program, it will attract people interested in sports and thus maintain its position. Accordingly, while people will move less in informational society than in current society, the movement will tend to accentuate the personalities of the various towns. If a town develops a distinct identity or personality, it will tend to attract people who like that identity and will tend to reinforce it.

Because people in informational society can attain prominent status through the social nervous system without leaving the town of birth, the incentives to move in informational society are much less than in current society. A person would only move to a town if he felt the town offered the activities and type of people he liked. Furthermore, as the town has the power to inhibit public expression of lifestyles it does not desire, a person would seldom choose to live in a town which frowned on his desired lifestyle.

Over time and through the mechanism of choice, most towns would come to be characterized by one or more compatible lifestyles. Because members of cosmopolitan towns would generally be more successful in national and international institutions, there would be forces in society promoting cosmopolitan diversity within a town. But, the fact that a successful town can promote only a subset of all the possible activities and lifestyles, places some bounds on the variation within a town. In contrast, individuals seeking alternative lifestyles would, through the mechanism of choice, tend to create diversity among towns within a metropolitan region. As people became less dependent on specific physical locations, there would, over time, come to be much greater diversity between towns than there is currently.

Lifestyles in Towns

The internal transportation system of a town would be designed to support the local lifestyle. Surface transportation in the metro area would be partitioned into the part controlled by the metrogovernment and the part controlled by the town. The transportation system on metropolitan land, for example on the greenbelts and other town boundary areas, would not interfere with town life in the towns; hence, the purpose in constructing this part of the metropolitan transportation system would be high speed transit between metropolitan centers and the entrances and exits of the towns. This transportation system would consist of some combination of roadways, subways and trains. The transportation system within the towns must compromise between two conflicting objectives; residents would desire rapid access to town activities as well as access to the metropolitan transportation system, but at the same time, travel within the towns should promote not hinder neighborhood life.

The conflict here is between speed of transit and the removal of strangers in towns and neighborhoods. For fast access and rapid transit times, the roadway system would resemble the current one-that is a grid of streets. For a society based on the automobile, rapid transit time promotes economic opportunities by increasing the distance a person can commute to a job or travel to shop. This, however, creates the problem of strangers or people whose only purpose in a town or neighborhood is to traverse the neighborhood. Too much outside traffic of this sort greatly deteriorates the quality of neighborhoods and contributes to the rise of property crime. As communication substitutes for travel, however, the demand for an intertown transportation system providing rapid transit in all directions will decrease, since, for example, information workers telecommuting to work and shopping through the social nervous system would not need rapid surface transit in order to exercise economic opportunity. Such a town could modify the transportation system to greatly reduce the problem of strangers. This objective could be achieved by incentives and the right of access. First, if the time required to bypass a town on the metropolitan roadways is much less than the time required to cross a town, individuals will have little incentive to enter a town without a specific purpose. Town transportation systems can accordingly be constructed to make traversing the town inefficient. Because they would have the power of access, the town could restrict access to sites within their boundaries. For example, a town might have only four exists and entrances and could have an access policy requiring outsiders to have the permission from at least one household to enter. Secondly, the local transportation system can also be arranged with *cul de sacs* to reduce traffic through neighborhoods[13]. With a decrease in the demand for rapid intertown transit, then, the transportation system can be modified to promote a sense of community and reduce crime committed by outsiders.

The extent to which a town would desire to modify the current land transportation system by reducing the number of its entrances and by modifying its streets would be dependent on how much the lifestyle of the town relied on current transportation practice. A town of blue-collar workers who required unrestricted access to get to work would desire a town transportation system that was similar to the one current in practice. A town of primarily information workers on the other hand, might wish to create a dual transportation system: one for access to activities outside the town and the other for activities within the town. The outside access transportation system would provide residents quick access to the metropolitan transportation system without encouraging across-town traffic. If all residences faced a town greenbelt, a secondary transportation system could be created for within-town travel. Residents could travel between points in the town on foot, by bicycle, or with special small cars. Older towns might modify their established system of roads to make them very inefficient for traversing the town and at the same time promote neighborhood life by blocking various streets to create numerous *cul de sacs*.

To illustrate how much internal transportation might differ from current practice, we might consider a town of information workers who perform work-related activities through the social nervous system. The town is organized so that the houses face a system of pathways. Since the town is a mile square, everyone can reach the town center in about 10 minutes. As this town places a high premium on human interactions, no bicycles or other vehicles are allowed on the pathway system. To transport physical goods and to provide an alternative when people would prefer not to walk a one way system of roads connects all houses in the town. To prevent outsiders from traversing the town, traffic is limited to delivery vehicles, bicycles, and electric town vehicles. Most residents own a road car which they must park in a garage at one of the four entrances to the town. This organization promotes human interaction at a minor cost in terms of the time to travel between two arbitrary points.

As people demand a wide tradeoff between human interaction and speed of access to town activities, there would be a wide range of alternative town transportation systems among towns. In physical arrangement towns would also differ greatly. Some would be organized so that most group physical activities would take place in a single village center. In others, group activities would take place primarily in neighborhood centers, and in a smaller number of towns most activities would take place in the home.

The common type of town in informational society would be an intergenerational community of people born in the town. Such towns might evolve from current blue-collar towns, professional towns, and ethnic neighborhoods in larger cities. One contributing factor to variation between towns would be wealth and income, and another would be the extent to which the town members substituted communication for travel. Almost all intergenerational towns would have sports programs, sponsor arts and crafts, promote social and belief events, and have a local economy. Residents would participate in numerous local events, would take part in other events though the social nervous system, and in varying degrees would leave the town for jobs and leisure activities. In short, each resident would pursue those activities which matched his talents and interests. In comparison with current metropolitan life, people would spend more time in local activities and less in driving. Additionally, with a decrease in the work week, most people would participate in more local social activities than currently. In a town which is religious these social activities might take place through the church, and in a secular town the golf club or other social club might be the focal point of social activities.

Most people growing up in a successful intergenerational town will have much less reason than currently for permanently relocating to another town. Information workers will not need to relocate for economic opportunity, as obtaining a new job will mean changing communication networks. Given a shorter workweek with fewer commutes, physical workers can commute further than currently. Two noneconomic reasons which would become more important factors in relocation would be to obtain support for a new lifestyle or to obtain better opportunities to pursue an avocation. These two reasons would tend to create specialized towns, since individuals who relocated to a town supporting a desired lifestyle would find the town organized to promote that lifestyle. While changes in wealth and income lifestyles would be common, a change in lifestyles based on belief would also become important. Successful towns specializing in a particular leisure activity, for example, artisan activities would be likely to be stable because they would attract from the general society those individuals who wished to pursue the chosen activity. The entrants to the specialized towns would in many cases be members of the intergenerational towns for whom pursuit of the specialty was more important to them than life in their childhood town. Specialties involving physical objects or activities would tend to be more concentrated in specialized towns than specialties involving mental processes, since a community of interest in a particular mental activity could communicate through the social nervous system. Most of the specialized groups in informational society would have a subgroup of members who would live outside the specialized towns and participate through the social nervous system and occasional visits.

In contrast to the intergenerational town would be the segmented-age-group towns. Towns which promoted an exciting night life and other single oriented activities would attract young singles wanting an exciting lifestyle. For those who were married or otherwise decided to have children there would be child oriented towns that provided strong support services to parents and children. Further specialization would lead to towns for very young children and towns for older children. This breakdown between age groups of children would depend on needs. Once children reached the age where they needed transportation, a town specialized for children who needed transportation to the town center would want to restrict access and provide a secondary transportation system to minimize the need for parents to continually drive their children from one activity to another. Finally, the elder citizens would retire in a more peaceful retirement town. The advantage of living in such specialized towns rather than in intergenerational towns is that a larger group with a particular interest can achieve economies of scale, and concerns about conflicts with other age groups would be minimized. As both intergenerational and age-group-segmentation orientations to life exist in current society they are likely to continue to exist in the future. Age group segmentation will increase given the need to provide the desired activities.

Besides speicializing in age-segmentation activities some towns would specialize in order to create a particular lifestyle or promote a certain belief. In the past two centuries numerous works have been written describing utopian visions of society. One famous contemporary work in this genre is B. F. Skinner's *Walden Two*[14]. This work describes a town whose organizational structure is based on Skinner's principle of operand conditioning. Currently a town, Twin Oaks, Virginia operates on a modified basis of Walden II. The attraction this town offers potential converts is the opportunity to participate in a richer community life than is found in contemporary society. The basic problem with the operating designs of Walden II and Twin Oaks is that they attempt to couple advanced concepts of psychology with a very primitive subsistence economy. In informational society, however, experimental towns such as Twin Oaks would have a much greater chance of success, since as the social nervous system advances and the length of the workweek decreases, the economic feasibility of such towns increases. Moreover, the social nervous system would enable such experimental towns to integrate their particular belief structures into organized society outside the town. While most towns that started with some utopian ideal would fail, they would nevertheless provide a mechanism for social experimentation. The more successful aspects of such towns could then be modified and utilized by more traditional towns. In short, as new technology makes alternative forms of social organization feasible, the utopian town provides the mechanism for social experimentation.

While some towns in the proposed design will specialize in experimenting with lifestyles of the future, others might well attempt to recreate the past. Already, in current society, there are weekend groups that model their activities on the activities of a medieval court, for example the Society for Creative Anachronisms. With greater amounts of leisure towns might arise that would attempt to recreate the lifestyles of previous centuries. For some of the towns these efforts would produce income sources, in the event that the towns become popular tourist attractions. Other individuals, like members of certain religious groups, would create particular forms of community out of conviction rather than as attempts to increase their wealth.

In informational society some towns would become more specialized simply because people would have more time to develop pursuits in activities which were formerly considered avocations. Take for instance hunting. In current society people usually do not have enormous blocks of time to devote to hunting. But given much more leisure time some people who like to hunt might want to increase the challenge by reviving hunting techniques of the stone-age hunter. As a result a great deal of energy would be taken up by learning (recreating) forgotten arts such as tracking. Hunting big game with stone-age spears would be enough to provide the hunter with a genuine thrill. A town which developed such skills would probably have a marginal income available to it through providing instruction to novices. Other specialized sports towns might exist for sports requiring major investments in equipment and facilities, for example motorcycle or automobile racing, polo, or rocketry.

While intergenerational towns are likely to have members participating in either arts or crafts, some towns would tend to specialize in one or a small group of the arts and crafts. Such towns would be organized to support the specialized activities. For example, in a pottery town, the town associations would provide various types of kilns and other pottery equipment. And since the main activities of this type of town are specialized, one would generally expect the level of skill to be very high. In such towns the specialty craft is also likely to offer some economic opportunities to the town members. The craft guild, for instance, could offer classes both live and through the social nervous system. In addition, fairs for artisan works could be held in the village center and a really successful artisan town might have shops offering wares at all times. Other specialized towns would arise promoting all of the arts and crafts.

Another type of specialized town would be the intellectual town. In such a town the principal activity might be the study of a particular subject, such as mathematics or chemistry. In a research town, research association funds together with grants would provide the basis for research in the desired subject. The research association members would obtain additional income from intellectual property, consulting and seminars. A great change from the 20th century will be the return of the leisure researcher of the 18th and 19th century, and groups like Mensa could have towns devoted to intellectual pursuits. A specialized intellectual town not of the sort specifically organized in todays society might be a town of Naderites, whose self-appointed function would be to analyze and criticize existing institutions. Such a town might compete for grants for such studies and might conduct some studies on its own authority. The importance of this function of continuous analysis of the systems performance will be emphasized throughout the remainder of the book.

In the field of business, specialized towns would arise for managers and entrepreneurs. With a constantly changing technology there will be no shortage of investment opportunities. For this reason, some towns could be devoted to full-time pursuit of financial investment activities. Business managers, as they assess their chances in the management situations market, might prefer a town of other managers where there was a considerable body of informal information on opportunities. Other business towns would become havens for entrepreneurs promoting startup companies in new technology.

Moreover, the growth of international trade will create incentives for the creation and maintenance of bilingual towns especially in the major metropolitan areas. In the past neighborhoods of emigrants generally lost their foreign culture and language as the subsequent generations were assimilated into mainstream US society. However, with the vast expansion in world trade second and succeeding generations would have an incentive to maintain both languages and cultures which will be supported by the availability of international media through the social nervous system. In addition, such bilingual towns would have economic incentives to maintain both languages and cultures in order to serve as language training centers for students in the metropolitan area. For example, town members might only speak the foreign language during such visits. Consequently, the number of bilingual towns will increase.

Local Activities

The function of a town, regardless of the degree of specialization, is to provide a framework for local activities. The local activities most frequently provided for local participation are sports, arts and crafts, social, and the local economy. Much more than current practice participants would use the social nervous system in local activities. First, the social nervous system will greatly reduce the cost of forming common interest groups for social, economic, and political activities. The initial search for common interests would be machine talking to machine. Second, people in towns would relate to outside institutions much more through communications such as teleconferencing. Because of the low transactions costs for organization and ease of interaction, many local activities would be integrated into regional and national organizations. Third, materials, such as educational software, would be available in the social nervous system to support local activities. Finally, local decision-making will increasingly involve analytic comparison of alternatives.

Consider sports and other physical exercises. Based on the assumption that the current trend towards physical fitness is a permanent shift in values in advanced countries, nearly all towns would feature a sports program which would provide activities for all age groups and all skill levels. In comparison with contemporary society, the proposed design would facilitate almost universal participation in sports or other physical exercise for both sexes and all age groups. Competitive sports would be organized into intertown leagues based on skill levels. In a metropolitan area with a large number of towns the leagues could be rearranged each year to maintain approximate parity of skill level. Most competitive sports would involve practice at the town facilities and travel to other facilities for games. As practice generally involves much more time than competition, the transportation system would well serve sports activities.

What would distinguish sports in contemporary society from their position in future society is not only the greater participation rate of all ages and sexes, but also the use of the social nervous system to promote higher levels of performance. The organization of sports programs would be conducted through the social nervous system in a variety of ways. These would include the obvious, such as record keeping and conference organization, but would also extend to the latest application of high technology to sports. As the reader may be aware, by attaching sensors to the athlete's body the computer can analyze the athletes performance[15]. In the future, these techniques will gradually filter down to the level of town sports programs. Thus, for instances, in teaching tennis to children, the instructor could have the child wear a sweat suit with the built-in sensors while he attempted to hit the ball. The computer program after analyzing the child's stroke would suggest improvements. Moreover, if sports were organized nationally, a small fee could fund continual improvement in performance through the creation of computerized aids to instruction of the latest techniques. With the social nervous system, superior athletes would be identified early, and in some sports, could receive professional coaching locally through teleconferencing.

Second, consider social activities. Most towns would have churches, areas for events such as concerts, and social clubs. In comparison with towns in contemporary society towns in the projected design would offer many more social activities for a variety of reasons. First, people will have increasing amounts of leisure to engage in nontraditional work activities. Second, many people find current society lonely[16], a fact which indicates a need for more social activities. Third, towns would remove the problem of strangers traversing the towns and at the same time arrange the transportation system to promote a friendlier atmosphere. And finally, as upward mobility would occur primarily through institutions in the social nervous system, individuals in order to achieve status, would not need to relocate as frequently as currently. Consequently, people in neighborhoods would socialize more than in current society. At the same time, considerable socializing would occur through the social nervous system. Currently many people visit using the telephone, but as the cost of teleconferencing falls, people will use it in their leisure activities as a new opportunity for conducting casual group meetings.

Third, consider skill activities such as arts and crafts. In industrial society people manipulate physical objects primarily to earn a living and secondarily as an avocation. In informational society the roles will be reversed. An individual might take up art simply because he found it intrinsically interesting and it provided him an opportunity to demonstrate a skill. To meet these sorts of interests most towns would have a variety of arts and craft programs. For example, a town might offer programs in carpentry, electronics, jewelry making, painting, and drama. While automated manufacturing will produce goods much more cheaply than hand manufacture, some people would want a few artisan objects in their living quarters. Like sports, these arts and crafts activities would be organized into national guilds through the social nervous system. The national guilds would provide a system of standards by which to judge the excellence of workmanship. Most arts and crafts would have a system of skill levels through which the devotee would advance to become a master. Guilds would also organize shows which could take place either through the social nervous system or at an actual physical location. The truly skilled in arts and crafts would earn some income through their efforts. Most, however, would participate to demonstrate competency as a test of self.

Towns in informational society are primarily residential as major economic activities would take place on metropolitan land. Nevertheless, most towns would have some economic activity. Already mentioned are successful arts and crafts activities. Beyond this, most towns would have a town center with perhaps a restaurant, bar, and a convenience store. In addition most towns would have some handyman type jobs in repairing equipment and yardwork. In households, most appliances and other equipment would be modular to facilitate easy maintenance by component replacement and would be monitored by the home computer to indicate when a replacement was required. While residents could replace the components themselves, people might not care to perform the wide variety of tasks this would entail. In addition towns with children would have services such as education and daycare. With specialization people would achieve higher incomes. To facilitate the town market most towns would have a market area for local economic activities.

Yet another activity which would be transformed by the social nervous system is local primary and secondary education. In current education the major conflict is between the power of the professional educator and the desire of parents and other citizens for more local control. For a variety of reasons such as the efficiency and diversity of larger schools, desegregation, and the increase of expert authority's power-in this case the professional administrator, the local control of education has greatly decreased. In order to financially support a diverse program of education to meet the needs of variations in student talent, students are currently being bused to large schools frequently outside their neighborhoods. The advance of the social nervous system, however, will facilitate the transfer of control over primary and secondary education back to the local community. Schools would be local neighborhood schools connected with a town high school, though the ways in which the educational system in a town was organized would vary widely. In some towns children would be educated in the home, while in others children would attend the traditional local public school. In still other towns there would be only private schools. The metropolitan government would offer specialized educational services through teleconferencing such as enrichment programs for the gifted, education of children with special learning difficulties, and expensive equipment such as laboratories. An increasing component of education will be computer-assisted instruction[17], and the development of this software in the social nervous system would be the responsibility of the federal and regional governments. At the local level teachers roles would gradually change from classroom lecturing to more individual counseling on finding the material in the social nervous system that is best suited to the individual student. As the quality of this material improved, the student could begin to obtain his education at his home terminal. Moreover, once the instructional material was in the social nervous system it could be used locally by any group. This would make possible the local organization of schools offering the latest technology of instruction through the social nervous system.

What clearly distinguishes future towns from present towns are the number of activities that could be pursued through the social nervous system. A resident would, for example, be able to access any book, film, or tape in society and in towns which developed a strong sense of community with similar values and tastes, the town would create its own programming sequence for television and stereo. An individual in the town could modify this sequence or, for that matter, any other programming sequence to create his own program sequence for his television. In addition, as the social nervous system advances the quality of computer games will also advance evolving ever more detailed, vivid animation and team games with complex roles. Using teleconferencing, an individual could participate at home in many activities which currently involve travel, such as bridge, chess or backgammon tournaments. The advantage of organizing such activities through the social nervous system is that, given a large pool of players, each individual would play against players at the same skill level.

The social nervous system would modify the primary manner in which people relate to organizations outside the town. In contemporary society the preferred type of meeting is a face to face meeting and telephone conversations are a secondary alternative. To achieve this end, great emphasis is placed on rapid transit between physical locations. This means that interactions with organizations outside the town generally require transit to particular physical locations. With the advance of fiber optics the face to face meeting becomes a teleconference and communication displaces travel as the primary manner in which individuals in towns interact with outside organizations. This means that organizations outside the town reside in the social nervous systems and are directly accessible from any terminal. Residents in a town will use the social nervous system to relate to leisure, economic, and political institutions outside the town.

The social nervous system creates the potential for low cost organization of new groups or new activities through machine talking with machine. For this potential to be realized, standards would have to be created for the format of queries, and in the creation of such a system, most household would desire software to screen incoming calls automatically. Once standards and the software had been created the formation of interest groups becomes the task of machine talking to machine. If a town member wished to organize a new activity, he could simply have his computer query his town to find potential participants, and, in turn, the potential participants would be informed of the activity by their personal computers. Two activities which are likely to be organized through the social nervous system are shopping and politics. Consider first something as simple as shopping for groceries. As computers become easier, to use it will be fairly common to use the computer to plan meals for such requirements as calorie control and to determine the food requirements for the family. In a town of people with similar tastes, the computer can also quickly aggregate all the purchases for any subgroup of the town so that the town could operate as a single purchaser. Having an inexpensive way to organize and manage buying groups will become more common as the social nervous system evolves. Their economic impact will be discussed in the next chapter.

In a similar fashion, the social nervous system will be used in political activities, by, for instance, making possible the inexpensive analysis of a politician's voting record and his actual positions. If a town is close-knit, it has incentives to sponsor a political analysis group for determining which politician really supports the town interests. This subject will be taken up further in a later chapter.

Beyond recreational, economic, and political activities, the advance of the social nervous system will enable residents of towns to make analytic comparisons of alternatives in a wide range of decisions. For the individual or family in informational society, a fundamental choice question is which town to reside in, since this particular choice determines both the activities they are likely to pursue and the type of lifestyle that will be supported by their neighbors. The decision, moreover, would be a complex one because the number of towns in a major metropolitan area would be in the thousands. Comparisons of this number of towns intuitively by any but the most superficial criteria would be very tedious, since to compare his alternatives analytically a potential resident would need to know for each alternative the quality and price of residence, the way the town is organized, the range of activities offered by associations within the town, and the types of lifestyles supported, tolerated, and discouraged by the town.

Currently, much of this sort of information is available in various information sources such as multiple listings of the real estate industry, voting records, the statistical abstract, and announcements of meetings of various activities. What is not presently available for the potential buyer, however, is a composite data base which would allow the buyer to consider his alternatives analytically at a reasonable cost. For example, suppose the sociology profession creates a test which is a reasonable predictor of the social compatibility of the buyer with the prospective town. A buyer might, then, wish to perform the test before making his choice of residence. For this purpose he would need an information right to obtain the input data. Such tests are likely to need information on the values held by town members which is not available in current society.

What makes this problem particularly interesting is that a family in a town relates first to its immediate neighborhood and then to the larger town. As friends generally have similar values, knowing whether one's values matched the values of the neighborhood would be of considerable interest. For this reason, there is little point to creating neighborhoods which promote neighborliness unless the neighbors are potential friends. Currently, however, in most residential areas the buyer only has the vaguest information concerning his future neighbors. This means that in order to make a wise choice the buyer would need information concerning the immediate neighborhood as well as the entire town.

The participants in buying, leasing, or renting a residence have incentives to use a database for comparing the alternatives. One group which has the incentive to create a database for all potential buyers is the real estate profession. To be sure, the members of a town would have mixed motives for providing information that would create a comprehensive data base organized for analysis, but the positive result of releasing the information, would be its role in attracting compatible new residents. The threatening aspect, however, is the danger that such information could be sold for other purposes, such as to create more effective advertisements or harassment. For towns and neighborhoods to release information, they would demand limits on the release of information. A real estate database company wishing to obtain a comprehensive database, might compromise by allowing potential buyers to use criteria to rank alternatives, without allowing the buyer to see sensitive raw data.

The problems involved in creating a data base in order that buyers of residences can make better decisions is one example of the importance of information policy, a recurring theme in this book. How accurately a person can assess his alternatives depends on the detail of information provided. If the real estate data base contained information concerning each neighborhood, the buyer could better find the neighborhood in which he was most likely to make friends. Such information could greatly reduce the search to find compatible people in the town and would accordingly make for more friendly neighborhoods. A buyer should be able to make analytical comparisons using those variables which established research indicates are important for the decision.

In information society the low cost of collecting and communicating information will always pose a great potential for abuse, no matter how much effort is taken to try to control the flows of information. To limit the damage from potential information misuse, each town should have the right to limit access both physically and through the communication channels. Most towns would have an access policy. If a town developed a strong sense of community wherein the members trusted each other, the town could create an access policy to become an effective buffer to the larger institutions in society. For example, as markets move to the social nervous system, sellers will have a detailed record of purchases made by individuals in a town. A town would be able to reduce this data to an aggregate for all town members by installing the town as an intermediary, using the town computer. The seller would have only a sales contract with the town and therefore would make a delivery to the town distribution center. The extent that the town would act as a buffer would depend on the extent that residents trusted their local institutions' ability and willingness to maintain privacy.

Individual versus Town

As the town in supporting a particular lifestyle is inhibiting or even precluding the public expression of alternative lifestyles, to what extent should the rights of the individual and his household be protected? For example, in order to promote family values, a town could suggest town households use software to filter violence and explicit sex out of social nervous system entertainment. In time, such filtering would be routine because to appeal to multiple lifestyles most new entertainment would have many, user controlled versions. The issue is should a town be able to compel an individual and his household to use such software when no other members of the town were present.

The need for individual rights depends on the extent every individual has freedom of location. Giving the town broad powers to promote the lifestyle of the majority encourages each individual to find a town which supports his or her chosen lifestyle. Suppose all individuals were completely free of location in the exercise of economic opportunity, that is they worked and purchased goods through the social nervous system. Given the diversity of towns, each individual could find a compatible town. Thus, total freedom of location would imply little need for individual rights.

The practical problem is that individuals will achieve freedom of location gradually and there will be great variation in the freedom of location of individuals in society. An information worker working through the social nervous system would have more choice than a blue-collar worker working a part time shift each day at a particular location. To protect the individual and the minority against the majority the power of the town to promote a particular lifestyle requires defined limits.

One important policy to protect the individual is the right of membership. Currently membership restrictions are usually in smaller human settlements than towns. One example is the singles only apartment complex. Another is the retirement community which restricts children. Most membership restrictions are implicit such as the wealth required to purchase a residence. In informational society towns could have membership restrictions which did not violate national policy. The basic policy would be the greater the freedom of location the more restrictions would be permitted. Restrictions based on race, creed or sex are very unlikely to become part of national policy. Some states might use court precedents to amplify restriction policies. The town would be obligated to provide evidence why a policy promoted the lifestyle while an individual could obtain an exception to live within reasonable commuting distance of a job. A member of a town could not be denied membership in town functions.

In current society the bill of rights and amendments grant the individual certain rights. Likewise the individual in informational society needs certain rights. A fundamental right of informational society is access to all of societies knowledge and information in the social nervous system. While the town to promote its lifestyle might wish to limit access to the social nervous system in its schools and might make suggestions the town has no right to restrict access or even to monitor access of the individual in his home. The power of the town to promote the town lifestyle stops at the residence of each family. At home each family has the right to pursue its own lifestyle in privacy. The individual has the right of freedom of assembly through teleconferencing through his home.

To promote town harmony the higher governments would have the power to place some restrictions on the power of a simple majority in town government. First on sensitive issues a higher than simple plurality vote could be required. The higher governments could vary this requirement as conditions warranted. Second the higher governments would determine the conditions for a town partition. If a substantial minority are fundamentally opposed to the policies of the majority and they live in definite neighborhoods in the town they can petition the higher governments for a partition. This check would prevent a 50.1 per cent majority from trying to impose its lifestyle on a 49.9 per cent minority without compromise. Other limits will be discussed in the chapter on government.

Evaluation

One feature of the metropolitan-town organization in informational society that would make life so much richer than in the current society would be the range of alternatives offered to the individual. Currently people in metropolitan regions primarily use travel by car to pursue the activities in their lifestyle. But as people are gradually freed from physical location, the choice mechanism, given the assumed social structure, would tend to sort people into clusters with similar lifestyle preferences, thereby freeing them from much of their present need to commute to and from recreational activities. Analytic comparisons of population characteristics recorded in town databases would enable the individual to quickly find possible matches.

The greater the freedom of location of an individual in his career, the more he can separate the institutional culture of his career from the chosen lifestyle of his leisure life. As will be discussed in subsequent chapters much of the institutional life of informational society will transpire via teleconferencing through terminals. An information worker, for instance, teleconferencing with workers in a corporation would maintain the decorum of the corporate culture while teleconferencing, but once he finished the teleconference, he could resume his leisure manners. If an individual had complete freedom of location he could mix any lifestyle with his role in institutional life.

Clustering people with similar lifestyles into small towns should lead to economic efficiencies. Through the social nervous system towns should achieve cost savings in purchasing, and with smaller groups with similar interests, the supply and demand for local public goods should be better matched. People in the new towns would also be able to pursue their chosen activities with considerably less transportation cost. Moreover, the social nervous system in combination with optional travel would provide the individual with more choices than currently. Finally, with control of random intrusion of strangers, the town should be a much safer and friendlier place than currently.

As a combination of market and political mechanisms would work to create the type of towns desired by the people, the town arrangement should be preferred to current arrangements. Through the choice mechanism the type of towns would respond to supply and demand conditions[18]. The towns which created successful lifestyles would be emulated, whereas towns promoting unsuccessful lifestyles would decrease. In existing towns such shifts would occur as the voting strength supporting various lifestyles shifted. Creating new towns would, of course, require the coordination of public and private interests. The metropolitan government would have to designate the land area for a new town, and the surrounding land would remain under direct control of the metropolitan government. Private developers creating new towns would tend to promote minor innovations with a forecasted market success; major innovations, on the other hand, are more likely to be created by groups committed to a new innovation. Also urban renewal of towns with old residences would require the political support of the town. For instance, if the land in a town became much more valuable in an alternative use, the town could vote to sell the entire town with a higher than plurality vote specified by the state government.

The design provides a natural institutional structure for social experimentation in lifestyles. As might be expected, most experiments in communes and other novel lifestyles will fail. Those ideas which succeed, would by diffusion, spread from the more experimental towns to the less experimental. As the pace of technology is likely to remain high, or even increase, new social possibilities will be made possible by the new technology. The choice mechanism provides a means whereby the social possibilities provided by new technology can be tested experimentally. Basing metropolitan organization on small towns would permit much greater variation than the current design, since current local political units are generally much too large to be responsive to variations in neighborhood desires. One point which requires experimentation is the optimal size of town. As people have differing needs for close interpersonal relationships, the optimal size of town would vary depending on the lifestyle.

As the workweek decreases towards zero and the cost of telecommunications falls relative to other goods, the lifestyles of towns could continue to diverge. In time, the diversity of towns could equal the diversity of now disappearing primitive cultures. Are there any desirable limits to diversity? The answer is yes, given the objective of continuing advancement of informational society which is is inconsistent with the generation of too many indolent lifestyles. For a variety of other reasons diversity must be limited too. For instance, given the kill capacity of the individual, violent lifestyles are undesirable. For peace to exist between towns there must be a common belief in a live-and-let-live attitude towards other towns. Since functions involving interaction between members of diverse towns will either be conducted through the social nervous system or on public land controlled by a metropolitan or higher government, there is no reason for random access to land controlled by towns. Each town has the right to control access to promote its own lifestyle. Furthermore, to promote peace each town has the right to practice its belief structure on the basis that it will allow other towns to practice their belief structure too. To be successful in organizations outside his own town, the individual would have to be tolerant of diversity. There are therefore incentives for a cosmopolitan attitude. As it is possible to live a reasonable full life within a town, this incentive for toleration does not specifically prohibit bigots, but it does mandate that they live within their own towns and leave other towns alone.

Notes and References 5

1. For example, see: Cunningham, H., 1980, *Leisure in the Industrial Revolution*, (Croom Helm: London) and Weaver, R. B., 1939, *Amusements and Sports in American Life*, (The University of Chicago Press: Chicago)

2. For example, see: Cheek, N. H. Jr. and W. R. Burch, Jr., 1976, *The Social Organization of Leisure in Human Society*, (Harper & Row: New York)

3. A reference for retirement villages in modern society is Osgood, N. J., 1982, *Senior settlers*, (Praeger: New York) and reference 1 contains descriptions of primitive societies which do little work.

4. Meyer, J.R., J.F. Kain and M Wohl, 1965, *The Urban Transportation Problem*, (Harvard University Press: Cambridge)

5. Meyer, J.R. & J.A. Gomez-Ibanez, 1981, *Autos, Transit, and Cities* (Harvard University Press: Cambridge)

6. For one of the seminal works on the use of urban design to reduce crime, see Newman, Oscar, 1972, *Defensible Space*, (The Macmillan Company: New York)

7. The author claims that current projections in the substitution communication for travel seriously underestimate the impact of the expansion in

bandwidth in communication and the impact of automation on increasing the ratio of information manipulation to physical manipulation of objects.

8. From 1960 to 1989 the number of residential community associations has increased from 5000 to 130,000 and about 12% of the population may now live in a residential community association. Residential community associations are private associations which provide services and act as quasi local governments with respect to such matters as zoning in condo complexes and tract housing. The rapid growth of these organizations is due to the economics of large scale development of residential housing. For a discussion see: ACIR, 1989, *Residential Community Associations: Private Governments in the Intergovernmental System* A-112, Advisory Commission on Intergovernmental Relations, (ACIR: Washington)

9. For example, see: Newman, Oscar, 1980, *Community of Interest* , (Anchor Press/Doubleday: Garden City)

10. The model of choice between the household and the town is the model of Tiebout. See: Tiebout, Charles, 1956, A Pure Theory of Public Expenditure, *Journal of Political Economy* , 64, Oct, pp 137-160. Also see: Zodrow, G. R., 1983, *Local Provision of Public Services: The Tiebout Model after Twenty-Five Year* , (Academic Press: New York). To obtain perfect mobility Tiebout fourth assumption is "Restrictions due to employment opportunities are not considered. It may be assumed all persons are living on dividend income." As informational society advances the substitution of communication for travel will enable households to consider a much wider range of alternative housing in relationship to work. Thus the fourth assumption would be approximately met.

11. Gans, Herbert J., 1967, *The Levittowners* , (Pantheon Books: New York)

12. One possible approach to creating more neighborhood oriented streets is to create private streets as has been done in St Louis. See reference 11 for a discussion.

13. Skinner, B., 1948, *Walden Two* , (Macmillan: New York)

14. The study of sports is a branch of biomechanics. Progress in biomechanics has lead to numerous models of sports behavior and simulation programs. See: Vaughan, C.L. (ed), 1989, *Biomechanics of Sport* (CRC press, Inc.: Boca Raton) or the *Journal of Biomechanics* and the *International Journal of Sports Biomechanics* In some cases software has been developed to aid the development of athletes. For example see: Persyn U., L. Tilborgh, D. Daley, V. Colman, D. Vijfvinkel and D. Verhetsel, 1986, Computerized Evaluation and Advice in Swimming, in Ungerechts B., K. Wilke and K. Reischle (ed), *Swimming Science V* , (Human Kinetics Books: Champaign) future advances should lead to a large number of software aids to developing athletes.

15. For example, see: Gordon, Suzanne, 1976, *Lonely in America* ,(Touchstone: New York)

16. Computer assisted instruction is just now in its infancy. While 90% of schools had at least one computer by 1985 so far CAI has had a limited im-

pact on instruction. For example see: Becker, Henry J., Using Computers for Instruction, 1987, *BYTE*, Feb, pp149-162. As advances are made in applying artificial intelligence to instruction, the impact of CAI will increase. For example, see: Self, John (ed), 1988, *Artificial Intelligence and Human Learning: Intelligent Computer Aided Instruction*, (Chapman and Hall: London). Also as the social nervous system evolves CAI will be available to all individuals throughout their lives.

17. In the US most new towns have been the creation of entrepreneurs. Ambitious federal plans to create new communities started in 1970 were suspended in 1975. See: Burby, R.J.III and S. F. Weiss, 1976, *New Communities U.S.A.*, (Lexington Books: Lexington)

Chapter 6

The Economy

Introduction

While the application of microbinics technology to production will result in a gradual displacement of people from the production of goods and services, the types of economic decisions which people must make will remain much the same. Households will have to decide how to spend their budgets. Investors must evaluate alternative assets. Managers, whether through computer analysis or intuition, must decide what products to produce. While the types of economic decisions to be made will remain unchanged, the process of making these decisions will be altered by technology which makes possible computer-supported decision making. The advances in the social sciences, the business disciplines and artificial intelligence are creating tools for analytical decisions and these tools are being programmed as software packages for use in decision making. Furthermore as hardware and software advance the cost of using these tools is rapidly falling. While at present they are used primarily in making business decisions, their use will gradually diffuse to many other types of decision making as well.

Decision making creates a derived operational demand for information; that is, data or observations are required as input to either intuitive judgment processes or analytical models. From a bounded rational economic perspective the decision maker will obtain more information up to the point that he judges the benefit exceeds the cost of obtaining it. As the analytic decision tools improve and processing and transmission costs of information decrease, the demand for information will increase. While advances in technology will increase the available supply of information, this supply is nevertheless limited by the incentives holders of information have for releasing the information. Sellers have positive incentives in releasing positive information concerning their goods and services; however, they have mixed incentives for releasing negative information. If negative information is likely to become public knowledge, the seller has an incentive to make a public relations disclosure to minimize the damage, but a seller has little incentive to release negative information which can be kept from the public.

Consequently, the supply of available information depends on information rights granted to those who make economic decisions. As will be discussed in the next chapter, the fundamental right in informational society is the right to learn. Based on this right, those who make operational decisions have the right to obtain information that will help them make decisions by analytically comparing the properties of the alternatives.

Many choices faced by individual decision makers are common to large numbers of decision makers. For such decisions economies of scale may be gained by creating databases for common use by all potential decision makers. One approach to creating these data bases that supply the operational demand for information is to treat operational information as a commodity. Because information is costly to acquire, a market perspective on information can reward entrepreneurs for creating information services that collect, process and sell information, provided the entrepreneur can establish property rights to ensure a return for his efforts. Much of the data that these information services would collect is data desired by decision makers to predict the performance of alternatives under consideration. Much of this data is generated by past behavior.

The advance of information technology and decision tools will understandably transform the consumer's approach to decisions. As communications advances to inexpensive dynamic image communication, most aspects of market transactions, with the exception of the delivery of physical goods, will shift within the social nervous system. Once data and analytical tools are sold by consumer information services, the consumer will be able to make analytical comparisons of his alternatives much more easily. With a high rate of new product introduction, the improved decision-making capability of consumers will identify the successful products much more quickly than current mechanisms can. Producers, accordingly, will need to respond to rapidly changing market conditions. As automation proceeds, the activities of the firm will increasingly become one-time adjustments to changing technology, product development, and market conditions. With much better information and decision-making capabilities managers will expand their use of markets to subcontract and hire temporary services. Managers will also employ communication networks to organize the firm's work as a sequence of tasks and will make greater use of markets to organize their tasks.

The driving force of change in the political economy is the government promotion of research. In the future funding for research to promote invention will remain at high levels and funding of research to promote innovation will increase. Part of the increase in funding to promote innovations will enable numerous experimental stations to experiment in alternative approaches to production and organization. Basic research to promote innovation must observe the political economy. Based of the right to learn the information policy for scientific study of behavior leading to innovations is at least an unlabeled representative sample. The representative sample is needed to accurately characterize the underlying behavior. As understanding behavior in general does not require labels identifying individuals, no labels should be provided to promote privacy. Improvements in translating new research into new products and innovations will take the form of improvements in the transmission mechanism for university research.

As larger firms rely more heavily on markets for subcontracting and temporary services, and as the number of startups and producer services promoting new innovations increases, more people will be employed by small to midsized firms. Work for all workers, whether managers, professional, information workers, or blue-collar workers will tend to become a sequence of unique tasks as routine tasks are programed. Because of the greater use of temporary organizations to accomplish these unique tasks, many workers' careers will be a succession of temporary contract work. Between contracts workers will retrain in preparation for new employment opportunities.

The Consumer

Consider the problem of the contemporary consumer. With an acceleration in the rate of discovery and invention, the consumer faces a difficult choice problem of evaluating an increasing rate of new product introductions. To evaluate his alternatives, the consumer has incentives to acquire information, and the seller has incentives to provide information.

The later, in promoting the adoption of new products, constantly invents new forms of advertising. From the perspective of the choice model, advertising provides information for three steps-generating objectives, determining alternatives, and evaluating alternatives. Advertising has an impact on the formation of tastes towards new products, makes the consumer aware of his alternatives and provides information such as price to evaluate the alternatives. Advertisers have incentives, in presenting material to consumers, to use the knowledge of the limitations in the consumers' cognitive skills to sell the product. They have no incentives to provide consumers with an impartial evaluation service.

Alternative sources of information for consumers are not well developed. With a rapid rate of new product introductions, knowledge through experience rapidly loses value. Information from word of mouth has no desirable properties from the perspective of sampling theory. For some items, such as personal computers, private magazines provide evaluation services. The best quality general evaluation service, Consumer Union, which publishes *Consumer Reports* only evaluates a fraction of the goods on the marketplace. The fact clients will give away information free to friends, inhibits private organizations from providing better evaluation services.

The government's role in providing evaluation services is to require numerous disclosures such as the ingredients in foods sold in supermarkets. However, the government frequently acts not to inform the consumer, but to protect him from any possible ignorance. For example, since the pure food and drug legislation of 1906, the food and drug administration, the FDA, has taken direct responsibility for testing drugs. The FDA releases drugs it considers safe. The consumer is not allowed to weigh the risk, unless he is willing to travel to another country with less strict laws.

Consumer performance will improve greatly with the creation of decision aids, advances in information technology, and an appropriate information policy. As technology advances in this direction, the locus of consumer purchasing will shift from physical stores to the social nervous system stores, with an increasing portion of purchasing activity taking place in the home. The extent to which consumers will be able to purchase goods through computer terminal devices in the household will be determined by the development of interactive image software which can be controlled by the consumer. In using this type of software for purchasing goods and services, the consumer will be able to identify and evaluate his alternatives. As consumers, on the basis of tastes and incomes can be stratified into large groups facing similar consumption decisions, there exist very large economies of scale to be exploited in the identification and analysis of alternative decisions. An appropriate information policy will enable entrepreneurs to innovate information services which supply consumers with software and data for making analytical comparisons of goods.

In contemporary society an increasing amount of household shopping takes place in the home through catalogs. When both adults in the family work, time for shopping is at a premium, and this creates an incentive for catalog shopping. With the advance of the home computer and the meshing of home computation with cable television, catalogs will become available through the home computer. The extent to which catalog in the social nervous system have more appeal than printed catalog or shopping in malls depends on advances in graphics, data organization, and interactive programs that give the consumer control over the flow of information. A desirable feature of computerized shopping is its ability to provide the consumer with desired, background information about products. Included might be complete specifications and visual simulations of operating characteristics. The reader might think that home shopping would never take place to clothing since most buyers like to try clothes on to see how they fit. But suppose buyers have a program which displays their wearing new clothes in many poses on their televisions. With such a program buyers could, through simulation, try on far more alternatives at home and in a far shorter time than trips to alternative stores would allow. Second, if clothes were ordered by a computer program which maintained the buyers' exact dimensions and were created automatically to fit those dimensions, they would fit far better than currently. In addition, the shopping program would arrange payment and shipment. Shopping for clothes in such an environment would offer buyers a far wider range of choices to consider and would enable buyers to evaluate their alternatives at a far lower cost. For example, alternative color combinations could be examined with the push of a button.

As the channel capacity of communications expands and computation costs fall, advances in such technology will reduce the cost of visually evaluating alternatives below the cost of onsite inspection. Moving the payment of purchases to a home computer function, however, requires creating software sufficiently "user friendly" that the average person is comfortable using it.

The shift of markets to the social nervous system will accelerate as the communication channel capacity expands to accommodate interactive image communication. Furthermore, the shift of goods to the social nervous system depends on entrepreneurs' ability to provide a superior environment for identifying and analyzing alternatives. Goods such as appliances will make the shift quickly because analytical comparison of their properties will be much superior to visual inspection without any efficient mechanism for comparisons. For products where other senses such as taste, touch, or smell are important, however, onsite inspection will remain important for some time. Shopping malls are therefore likely to become much more specialized in goods for which taste, touch, and smell are important and entertainment activities are likely to replace stores whose markets have shifted to the social nervous system.

In the current market organization the consumer encounters high transactions costs in identifying and analyzing his alternatives[1]. At present, price comparisons of alternative goods can only be made by examining alternative catalogs, traveling from store to store, or by phoning the alternative stores. Catalogs are not organized for the convenience of price and product comparison by the consumer, and travel is expensive both in time and resources. Price comparison by phone is frequently frustrated by the difficulty of precisely identifying the product being considered and the reluctance of sellers to give out this information over the phone. But for the consumer to take full advantage of the potential advances in hardware and software to lower his transactions costs, he needs certain information rights. His home computer must have access to product codes, to specifications of the products' qualities, and to the price offered by all potential sellers. But such a system of information rights would require a major legislative initiative to institute. Such an effort would be hampered by lobbyists for the inefficient would try hard to prevent the passage of such legislation.

The purpose of the right to learn is to categorize information such as that listed above as operational information needed to analyze alternatives. Under this arrangement, if legislation was not forthcoming, the necessary information rights could be defined by legal precedent in interpreting the right to learn. Once these operational information rights were in place, sellers who could not list their wares in the social nervous system would be at a distinct disadvantage. Accordingly, the focus would shift from ensuring that buyers had access to information to ensuring that all potential sellers had access to the electronic marketplace.

The ability of the buyer to obtain the price and qualities of the product information by having his home computer communicate directly with the computer of the potential seller creates the basis for improving consumer performance. First, administrative programs could organize the acquired data in order to analyze the relevant alternatives. For any given product the buyer could determine his best deal at a very low cost. And since the product information would be obtainable, the shopper could make analytical comparisons of alternative products by using analysis packages at home. This means that consumer could base household purchases more on analysis and less on psychological appeals of advertising.

Great economies of scale exist in exercising these information rights. Desirable data organization from the perspective of the buyer is quite different from optimal data organization from the perspective of the seller. Suppose the buyer wants to find the lowest price for a particular item. Under data organized for the benefit of each individual store, the buyer must access the database of each store, figure out how the data is organized, find the price, and finally obtain the lowest price. Each buyer seeking the same information must repeat the same steps. From the perspective of the buyer, all the data from every store should be reorganized by product in a single database. If the prices for each store were arranged in ascending order, the search for the lowest price would be reduced to single table lookup. Given property rights to processed information entrepreneurs would create consumer service companies to perform frequently employed searches once and organize the data for sale to the consumer.

To be a much more effective shopper, however, the consumer needs much more than a low-cost approach to identifying his alternatives. He must also have a low-cost method of evaluating their merits. In light of rapidly advancing technology, experience with past products is not a good guide if the next purchase involves newly designed goods. To provide a contrast with current conditions, consider first the problem of buying a horse in the 19th century. At that time the buyer accumulated information concerning the quality of horses over his entire lifetime. As the characteristics of horses changed very slowly due to breeding, knowledge from previous purchases could improve the buyer's future judgment. Contrast this consumer purchase problem with the problem of buying an automobile in the 20th century. Currently the technology of an automobile changes at a sufficiently fast pace that past information is of questionable value in making current purchases.

Advocates of pure competition assert that the market in its operation creates the information to evaluate the alternatives. Market information is generated and used by consumers in evaluating their alternatives. The question, however, is how efficiently the consumer obtains the information and how well he uses it. Consider the consumer versus a quality engineer evaluating the quality of a product. The quality engineer uses a statistically designed experiment to analyze the quality of the product being considered, whereas the consumer has only word-of-mouth information and information from trade magazines such as *Consumer Reports*. The problem with word of mouth is that it is does not provide a representative sample. For goods which are used by only a fraction of the population, then, word of mouth may provide little information. Furthermore, when it comes to highly technical products the consumer, for the most part, is not capable of evaluating the quality of the product. For example, most consumers are not capable of evaluating the issue of radiation leakage from microwave ovens, as the damage is not directly apparent to the senses. Furthermore, market competition is unlikely to reveal much useful information concerning leakage. The limited scope of existing consumer services are inadequate for providing consumers with all the product information they need.

To determine what information rights are desirable for evaluating the quality of alternative goods on the market, suppose that every time the consumer had a product repaired he received a machine-readable bill; that is, his home computer could access the information and store it. If the repair records of most available equipment were kept in a machine-readable form, systematic analysis of the quality of alternative products would become a straightforward matter. Assuring the consumer an information right to all his transactions, establishes the basis for a mechanism for better evaluating products. Again, there are tremendous economies of scale in the exercise of this information right. Given property rights over processed information, customers of consumer service companies would transmit their repair files to the consumer service company for statistical processing. Knowledge concerning product reliability would thus accumulate quickly and be available to potential buyers within the product cycle. Similarly, if accident reports were transmitted to the consumer service company, the statistical determination of the safety of alternative products is a straightforward matter.

The economies of scale in providing consumers with the basis for analysis will be obtained by co-operative consumer groups, private consumer service companies selling services, or by government consumer services providing free information as a public good. To obtain the economies of scale in evaluating products, these organizations would need to serve a large number of consumers. The problem with consumer services as they exist today is the high cost of performing tests and obtaining survey data. For example, in obtaining information on automobiles, the Consumer Union relies on a survey of its readers. As filling out this survey constitutes a burden to consumers, the participation rate is less than complete, and the results are unlikely to represent an unbiased sample. With machine readable repair records, however, the information would be available to any consumer service to which the consumer subscribed at almost no labor cost to the consumer. Sample survey questions would be limited to preference questions concerning such issues as liked and disliked features of a product. The burden of participation in such surveys could be greatly reduced by interactive programs and by making a fixed amount of participation in surveys a condition of membership. With complete information on the performance of their client products, the detection of problems would be much, faster than is possible through word of mouth. The consumer would obtain this information through his buyer services.

The biggest impediment to the growth of a strong consumer analysis industry in the current market organization is the free rider problem. If Consumer Union publishes an edition of *Consumer Reports*, any valuable information quickly becomes public knowledge without Consumer Union receiving any return on its investment in providing the information. Under the proposed design, however, consumer services would generate most of their income through charges for access time to run consumer software services. As the charges would be based on access time, the consumer service would collect even if a client allowed a friend to use the software provided. Suppose, for instances, a household wanted to buy a refrigerator. As this type of a purchase is generally made at very infrequent intervals, the consumer, when he starts his search, has very little information concerning the alternatives in the market. Furthermore, his neighbors are unlikely to have any information either, given the change in technology and the inability of the average individual to correctly assess the advertising claims. The individual, therefore, frequently relies on the image of the company providing the product and the salesman's personality.

However, suppose the consumer service offered a user-friendly program for assessing the alternatives in the marketplace. The program would start by finding out the user's needs through a series of questions. For example, if the consumer's grocery transactions were organized in a form accessible to the consumer service company, the demand for refrigerator space could be determined primarily by the consumer service company's communicating directly with the consumer's home computer. Then the program might suggest some alternative configurations and narrow the selection down to a few types. Next, the program would have at its disposal all the latest engineering assessments of the quality of all the alternatives in the marketplace. The consumer would then indicate some preference as to brands or specific options.

The program now sets up the relevant market from the perspective of the consumer. The consumer is asked the area to search-for example, the metropolitan region, the entire state, or nation or world. Again this could be made automatic. After examining the alternatives, the program displays the best, say 10, alternatives, and the best, say 10, alternatives in each of the close substitute markets as well. Now the question arises with his consumer service program: Can the consumer make a more informed judgment than currently? First, seated in front of his computer screen the consumer can review all the background information which went into his decision. He can display the product, review the analytical comparisons between the close substitutes, and search over all possible purchase plans in the market. His consumer service from the repair records can estimate the quality of service each of the alternative sellers is likely to provide. His program can recommend between the close substitutes based on the criteria he supplied. He can modify his criteria and immediately examine the consequences of his choice. His great advantage over present consumers is his ability to positively identify all alternative products and their qualities. In a physical search in today's market, it is hard to remember the qualities of the close substitutes, especially if there are a large number of similar products, and so the price quality comparison is very difficult.

Periodically, demands are made for government to take a stronger role in promoting the consumer side of the market. In the proposed design, the government first enforces the information rights of the various parties in the marketplace. When these information rights are in an electronic medium for information, powerful incentives are created for consumer service companies. Through surveys of their members and analysis of repair records, consumer service companies can use statistical techniques to generate much more accurate assessment of product qualities much more rapidly than current word of mouth and limited testing means allow. By creating software services that are either sold or run on the consumer service computer, the consumer service company can greatly reduce the free rider problem which currently limits the scope of contemporary consumer services.

This approach of granting consumers specific information rights contrasts greatly with the current trend whereby the government generates the information that is provided to the consumer. The amount of analysis the consumer can do in comparing his alternatives under this system is dependent upon the extent of his information rights. Consider the purchase of previously owned objects such as automobiles or houses. If the consumer's information rights extended to all previous transactions, then the consumer could analyze the status of the object objectively. Information rights can also improve the safety aspects of products. Suppose the consumer were granted the information right to inspect the product development and safety testing reports. Having this information right would inhibit corporations from knowingly releasing defective products. Under such conditions, it is improbable, for instance, that Ford would have released the Pinto, which posed a fire hazard if hit in the rear. Furthermore, with the complete report of the development of the product, the consumer services and trade magazines could forecast expected performance with some reliability.

Finally, information policy through market action would greatly reduce the need for direct government involvement in consumer affairs. As market firms, consumer services would offer a menu of services to attract customers. Some, in order to attract customers, would take an aggressive attitude towards promoting consumer's information rights in the marketplace. While individuals would seldom have the incentives to use the courts to obtain information, a consumer service with a million customers would. Much like current consumer rating services, the consumer software services would in varying degrees maintain independence from producers. Most of the positive impacts of information policy, in fact, would be achieved through the high quality services' efforts to maintain strict independence from suppliers.

Nevertheless, the federal government should determine the qualities of products for which the expense of testing and the economies of scale are very large as a public service. Currently the government provides a considerable amount of consumer information as a public good, but in the future some of these services may be taken over by consumer service firms. One example is the data on the gasoline mileage ratings of automobiles; if the consumer service were to provide this information, the clients automobiles on the road would record their actual gas mileage and the consumer service would report the tabulated results to all members. On measurements of quality for which consumer service firms have low cost methods of measuring the quality, there is no need for the government to provide the service. One area where the federal government has a legitimate role in consumer services are areas where the economies of scale are very great. For example, testing drugs requires both expensive laboratories and a highly paid professional staff. A second such area is the promotion of research into new methodologies for product testing and safety evaluations. In such cases new scientific theories must be translated into new test procedures. The government has a legitimate role in such instances of basic research.

If consumers obtain the information rights suggested in this chapter, their ability to identify and analyze their alternatives will greatly increase. Such a decision environment would appear to eliminate the contemporary role of advertising. With his consumer service the buyer would be much less susceptible to sales pitches based solely on psychological appeals, nevertheless a role for advertising remains. In the choice process there is one step which requires imagination: the manufacturer or service supplier can specify the properties of a product, but the consumer must decide how he plans to use the product. While decisions made concerning this step are obvious with familiar products, when it applies to new products this step taxes the imagination of many potential customers. To promote new products advertisers would therefore focus on bringing to the potential consumer's attention the innovations the product provides the user. Most consumer services would have a policy for including this type of advertisement, provided it met the consumer service standards, as background material which the client could view at his discretion.

Markets

Shoppers will be able to capture economies of scale much more easily, since the social nervous system facilitates much greater organization on the household demand side of the market. Consider the advantage of buying in large quantities to obtain quantity discounts. Where there is a sizable group of people with similar tastes in a town, they have an incentive to shop as a group. Within the social nervous system, home computers in the respective households will be linked by a network, so the total purchases of the group can be easily aggregated. For example, if the various households use software to assist them with dieting and food preparation, their home computers would maintain their grocery lists automatically.

As people have more leisure time, they will have more time to obtain the maximum value for their income. Thus in most communities, households which have similar tastes and live near each other are likely to form buying groups for commonly consumed items. The size of buying groups is likely to remain limited as with increasing size, the similarity of tastes would decrease and the costs of distributing the group purchases would increase. In current society there exist buying groups for such items as produce. In the future, however, these groups will jointly purchase a much more extensive list of products. A group will give its members access to more expensive equipment, for example a high quality hologram, a three dimensional image projector, to evaluate how a product might look in a home. The consumer group allows the group to obtain quantity discounts and at the same time obtain a reduction in the distribution costs, as the total order can be delivered to the community distribution center. Thus, the economics and the possibility of low-cost administration of group buying and delivery will provide the impetus for the development of buying groups. In part, the savings come from eliminating the retail display costs and from the fact that the buying group absorbs some of the distribution costs. Because of efficiencies made possible to the buying group through the social nervous system, the total labor costs of each member of the group in shopping will decrease.

Buying groups making their purchases through computerized markets will profoundly alter the mechanism for distributing goods from the factory to the household. Many items will be bought directly from the factory, with the manufacturers' computer aggregating orders for efficient shipment. Since the buying groups in the communities can easily aggregate orders, their purchases would be shipped in a single order and distributed automatically when they arrive at the town distribution center. To the extent that they desire to have products without delay, consumers will be willing to pay more to reduce the delay. In many cases they will prefer to buy from a local distributor at a slightly higher price than than obtaining the product from the closest factory. Depending on the nature of the market, there might be a large- and a small-order market, necessitating a distribution system such as wholesalers and retailers. Because the cost of determining the lowest price will be greatly reduced by information rights and increasing computational power, the number of middlemen would decline in most markets. Two factors will greatly reduce the number of retailers: First, price competition will eliminate the inefficient, and second, retailers will face a free rider problem. Retailers display goods so that consumers can make onsite comparisons. The overhead for such displays must be incorporated into the prices retailers charge their customers. With computerized markets, however, retailers will find that potential customers will increasingly use retail establishments primarily to examine alternative goods. Once they have selected a good for purchase, the consumer will then find the best price through the computerized market. Warehouses without displays will therefore be more price competitive, and as the consumer can obtain better product information through the social nervous system than from any sales person, the value of sales persons also decreases to the consumer. Local stores will thus tend to become automated warehouses with goods for the metropolitan market. The trend would be for manufacturers to permanently display their wares in display centers or traveling displays. The manufacturer might have a single display center for a metropolitan area, and traveling fairs featuring new products and demonstrations of new concepts will make a resurgence to replace the services formerly provided by the disappearing retailer. The role of these live displays will be to supplement consumer information services.

As markets move into the social nervous system they will generally acquire a more dynamic price adjustment mechanism for equating supply and demand. One important characteristic of such markets will be their capacity to make price adjustments either on line or off line. In most stores today prices are adjusted "off line" by the seller, and the buyer faces a fixed price. For some goods such as cars or houses, however, the buyer and seller adjust the price online in a bargaining process. As the construction of a computer program for on-line adjustment of prices would be considerably more expensive than an offline mechanism, most consumer goods would be sold with fixed prices. But for some goods the computer market would be an electronic version of the pits of the Chicago Board of Trade Commodity markets. Markets of this type would initially be for more expensive items such as automobiles or houses, where price adjustments in the selling prices are common. But as microbinics costs fall and as the growth of buying groups for volume quantities of goods become more common, one would anticipate the growth of on-line price adjustment markets to provide suppliers with a mechanism to compete for larger orders. Computerized markets could provide complete market information to all participants on computer screens. The price setting mechanism of such a market would be the instantaneous interaction of supply and demand, with speculators smoothing out the fluctuations. The growth of such markets would evolve from the software now being developed for computerized stock markets. Given more discriminating consumers, competition between similar products for sales in a market with consumer information rights would quickly establish price differentials reflecting quality differentials.

In response to more discriminating customers, firms will have to modify their strategies. With a high rate of discovery, invention, and innovation, firms will introduce a constant stream of new products into the marketplace. With better evaluation procedures, consumers will more quickly determine the winners creating sudden and frequent shifts in demand. Moreover, with a higher rate of discovery and consumer information rights, unsuspected negative aspects of products and services will be discovered more quickly. To compete in the marketplace firms will need the ability to rapidly adapt to reduce the risks in introducing new products. To reduce this risk, firms would design production facilities with the capacity to produce multiple products and to rapidly shift production in response to demand shifts. Where possible, firms would rapidly improve their products to possess the attributes of the winners when analyzed by potential customers. Firms would purchase the information describing the attributes of winners from the consumer services. Provided that individual consumer behavior was not revealed, consumers themselves would sanction this sale, since it would reduce the cost of the consumer service to the consumer.

Organization of the Firm

To compete in rapidly changing markets, firms must develop mechanisms for organizing work that efficiently consider the impact of automation on the nature of work. As was discussed in the previous chapter, the gradual programing of tasks in a firm will leave a sequence of one-time tasks to be performed by humans. Management will focus their attention on the problems of the sequence of inventions and innovations. With the routine aspects of staff work encapsulated into software such as expert systems, human staff tasks will become more random and will require a wide range of skills. With market transactions taking place through the social nervous system, the sales force and customer representatives will handle the exceptional cases. Workers in manufacturing plants will increasingly handle nonroutine events such as installing new machinery or repairs when machinery malfunctions. Given the nature of work, firms will need much more flexible organizational structures than is custom currently.

Instead of requiring a stable work force performing primarily repetitious tasks, firms will need a variable workforce which can change in response to the demands of the one time projects to be accomplished. Recent trends towards establishing such flexibility[2] have been the increasing use of temps, or temporary workers, and the trend towards transferring activities previously performed in house to activities contracted in markets.

As the firm's activities become a sequence of one-time tasks the use of temps and outside contracting should continue to increase[3]. For example, because the firm would need temporary staff services possessing special skills at random intervals, the firm could obtain the service best suited for their various needs by hiring producer services on a temporary basis. In this way, given an efficient market, firms would achieve greater efficiency than would be possible if they maintained a large staff of specialized producer services that would only be needed at random intervals. The efficiency of producer services should increase as many of these services could be supplied through the social nervous system by means of interactive visual communication. As markets in the social nervous system expand in scope and as competitors become less dependent on physical location, the number of participants competing for any opportunity should increase.

The efficiency with which a firm can use markets to organize tasks depends on the efficiency of markets. More efficient markets will result if both buyers and sellers have better means of identifying and evaluating their alternatives. As markets move into the social nervous system, the producer will be able to identify all his alternatives quickly by computer. The problem which must be overcome to make such markets efficient, however, is developing analytic techniques capable of evaluating the large numbers of alternatives.

The difficulty in developing analytic techniques to evaluate alternatives depends on the uniformity of alternatives. To aid the producer in purchasing uniform inputs, such as office equipment, economic incentives would create producer buying services to provide evaluation criterion. Large firms, which have economies of scale in testing the quality of their own purchases, would have an incentive to recoup part of their testing costs by selling the information to the producer purchasing service.

The more difficult problem remaining, then, is the task of providing information and criteria for evaluating nonuniform situations. Consider, for instances, the problem of a manufacturer subcontracting the production of a part. To evaluate the alternatives analytically the producer needs to know much more than just the prices bid by the competitors, since the producer's profits are a function of the subcontractor's ability to deliver the part on time and without defects. To evaluate the bids analytically, then, the producer would have to have a database with all the requisite information. For example, in order to evaluate analytically a large number of advertisers who might develop a promotion for a new product, the producer would need analytical measures of past performance.

Producers face similar problems in output markets as well as input markets. One example is the case of a bank providing credit for customers. The bank must evaluate the risk of default on a case-by-case basis. For this task, credit rating services have evolved which provide banks with measures of past credit performance.

The extent that information services will evolve to provide analytical information and criteria depends on the capability of advancing technology to provide useful measures. With the advance of microbinics, performance in the workplace will increasingly be measured numerically and, accordingly, the ability to make more detailed measurements will also advance. Information services providing information necessary for evaluating alternatives in most markets will flourish. For example, for subcontractors, the accumulation of data on on-time delivery and number of defects becomes automatic. Developing an information service to evaluate subcontractors in an industry requires major contractors establish conventions to collect the data in a comparable form. In advertising, the effectiveness of promotions will be measured in terms of customer response. Establishing conventions establishes the basis for an information service.

The growth of information services will enable firms to efficiently screen large numbers of external alternatives, such as temps, subcontractors, or producer services, for one time tasks. Suppose a firm needs to hire a software service to develop a new program. With teleconferencing, the relevant market for software services is worldwide. When the firm places a request for a proposal in the social nervous system, the firm is likely to receive a large number of bids. Because of cognitive limitations, the firm could not, in a meaningful fashion, screen a large number of bids. The firm would be forced to use an arbitrary criterion to reduce the number to an intuitively manageable level. With performance files provided by an information service, the firm would be able to develop meaningful analytic criteria to screen all the bids quickly to obtain a small number of promising candidates for a more intensive intuitive screening. The advantage of the initial, analytic screening over an initial, subjective screening is that the analytic approach is much more likely to reveal up-and-coming firms.

The discussion, so far, has presented three reasons why firms will increase the use of the market in organizing tasks. First, automation will reduce work to a sequence of unique events demanding random special skills. Second, competition will be worldwide because many of these skills will be available through the social nervous system. Third, with much better information and analytic capability to analyze alternatives, the firm will be better able to match its needs in the market. An additional reason for firms to use the market in organizing tasks is to reduce risk. Firms will want to subcontract secondary activities undergoing rapid technological change not directly related to the core of the firm's business. Given the bounded rationality of managers, the ability of the manager to administer the core activities will be increased by his not having to focus on rapid change in secondary activities. With more efficient markets, then, secondary activities in the firm will be increasingly subcontracted.

The more efficient markets become, the more a firm could institutionalize change by transforming its business into a sequence of contracts through the market[4]. With each renewal of fixed-period contracts, the structure of the firm would be reorganized to best meet the current challenges. At the extreme limits of this scenario the firm becomes a network organization which contracts out most of its activities; the firm therefore exists as a set of contracts. The extent to which a firm becomes a network organization depends, however, on the conditions in an industry. One factor limiting the extent of contracting is the uncertainty of supply, since a firm would prefer to maintain an activity in-house if the supply were uncertain. Another limit of contracting is the fact that a firm would not contract activities of proprietary concern. The more the performance of activities in a firm depends on close cooperation, the less incentive a firm has to use markets for organization. Such firms would tend, instead, to focused on the core activities of their main product lines.

In addition to a greater reliance on external markets, firms are also likely to use internal markets more extensively to allocate resources[5]. Because of their humanly limited cognitive skills, managers can effectively manage only a small number of subordinates. With analytic management based on measured results the number of subordinates a manager can effectively manage will increase. Firms with multiproducts will therefore have fewer managers in total, but each manager will manage more subordinates. Product managers, for instances, would compete for production runs on flexible manufacturing systems or computer resources. As a result of larger numbers of managers and temporary groups competing for internal resources at each level, firms are likely to increase the use of internal markets to allocate resources.

To achieve greater flexibility firms would depend less on permanent hierarchy and more on temporary task organizations. As informational society evolves, the amount of information available to managers for decision making will increase. At the same time, the number of levels of middle managers needed to process the raw data into reports for higher levels of managers will greatly decrease. Using programs such as expert systems and report generators, higher managers supported by a few assistants, will analyze data bases themselves. Hierarchies in firms such as corporations will thus have fewer levels of managers. Consider, for a moment, the role of the manager within the projected system. Within his firm a manager would spend much time using the social nervous system forming and directing direct temporary organizations to accomplish specific tasks. Most of these task organizations will be accomplished by establishing a visual communication channel between the members of the taskforce. In this manner a firm can very quickly bring the appropriate people in contact with one another to solve a particular problem. A firm would also have the ability to reconstruct communication links on demand. In other words the organization of a firm would be the network of communication channels linking the members together. Within such an organization, the physical location of the members of the various taskforces would decrease in importance. This would enable task organizations to include members of the firm and along with members of subcontractors or producer services. As understanding of how to organize activities through dynamic image communications increases, more of the permanent activities of organizations will take place through this medium.

In addition to relying on frequent temporary task reorganizations, the underlying organizational structure will require reorganization on a regular basis in order to incorporate new knowledge and technology into the organization. Given a rapid advance in knowledge, firms will have a problem in matching the most qualified manager with new openings in the management hierarchy. The solution proposed here is the creation of a management situations market of finite time management contracts. A high rate of scientific and technological advance will force managers to devote time to intensive study of advances which are applicable to their industry. Because of competitive pressures on the job, managers will not share the general decline in the workweek experienced by other workers. While on the job most managers will be working too many hours on projects to have any time for significant study. Consequently, managers on the job would be using primarily acquired skills. Upon completion of their contract, however, they would take a vacation and engage themselves in a period of study before reentering the management situations market. The advantage which the management situations market gives the firm is that it institutionalizes the need for constant reorganization and evaluation. With each new series of contracts, the firm can make substantial changes in organization, technology or product line. As managers would take a break between assignments, the not so humorous problems of the Peter Principle would be avoided. And as the substitution of communication for travel advances, the manager like other professionals, would have increasing freedom of location. What this means is that for any management position the number of potential managers who could assume the role without physically relocating their home will gradually increase. With automation of middle management the number of management positions will decrease; consequently, there will be an excess of potential managers as compared with the available management positions. Since the function of management can take place through the terminal, in the long run the potential competition for each available management position is worldwide.

From the perspective of the manager, the management situations market gives the manager total control over his career, since he can select the sequence of contracts upon which he wishes to bid. For a management situations market to work much better performance data would be necessary in order to evaluate managers. With increasing numerical measures, each manager would have a performance history in terms of such measures as the profitability of previous profit centers managed and the number of successful new products introduced. And with the right to learn, information services would create databases to facilitate analytical comparisons of the past performances of managers.

The competition for line manager positions would be a two-stage process involving a preliminary analytic screen followed by a more intensive evaluation of the final candidates. The details would vary by industry and the firm. Consider, for a moment, a possible scenario. First, the corporate or higher level of management would employ analytical criteria based on past performance to screen all applicants in order to obtain a small group of final contestants. The evaluation criterion could use any measure defined over the applicants' accumulated performance. In most cases the evaluation functions would be a weighted sum of a variety of criteria. A service industry would develop to make recommendations on how to specify the criteria to achieve specified objectives. The criteria to meet social objectives such as no discrimination would need to be substantiated by systematic study.

The second stage would be a more subjective evaluation of future projected performance. One such approach to final selection of a new manager is a competition through management proposals. The final contestants might be given a thirty-day contract to prepare their proposal on how to manage the prospective operation. The evaluating managers would establish the operating guidelines, such as the amount of investment money allowed for improvements, the goals, such as cost reduction or increase in the share of the market, and the conditions, such the decisions which are decentralized and which must be cleared through higher management. The bidders might also be provided with the proposals of outside management consultants. The objective of the evaluating manager, then, is to ensure through competition of managers that all profit centers of the corporation obtain their highest rate of return. The evaluating manager could specify whatever criterion he desired for the profit center. The objectives could be, for instance, some weighted average of profits, return on assets, growth, and share of the market. An alternative would be to simply have the potential managers compete to lease the profit center or where the production facility is producing a component of a larger durable good, the criterion might be cost minimization.

The final contestants would carefully study the current operation and decide how to apply new knowledge and technology to improve performance. Their proposals might include recommendations for purchasing certain types of technology or software, suggestions about how to reorganize lower levels of the organization under their control, and proposals for new product strategies. One approach to the bidding process might be to have potential managers compete as management teams, that is small groups with the requisite diversity of skills to manage the contract for which they are competing. Some of the management teams would form while the participants were in their twenties and would continue throughout their careers. Relationships within the group would therefore come to be personal relationships and the internal organization would vary from group to group. Some would have a clearly identifiable leader, while others would have an inner circle of approximately equal partners.

The competition by management proposal or bid ensures that new knowledge is incorporated into the firm. The managers of mature industries in informational society, like their current counterparts would be professional managers meaning that most of these people will include in their training formal study in management techniques such as those taught in an MBA program. What is different between future and current managers, however, is that a management career would have the quality of a work-study approach to education except that the work-study period would extend through a manager's entire career. During the period that managers were not managing they would be vacationing and then preparing for the next contract. To compete they would have to study intensely the latest advances in knowledge and technology applicable to their industry. The management proposal would be the means by which they would apply their knowledge to a particular situation.

With a constant cycle of fixed-period contracts the structure of a corporation would adapt to changing conditions with each round of competition for positions. This process of change would emanate downward from the corporate management group, the management group at the top of a corporation's hierarchy. The functions of this group would be to manage the day-to-day operations of the corporation and to make investment decisions in software and hardware, including the more traditional plant and equipment. This group would manage the day-to-day operations of the firm through the criteria specified in the management contracts of lower-level managers. The investment decisions would include expansion through merger or creation of a new plant and restructuring through sell-offs and spin-offs. Most major investment in research and development would be decided by the corporate managers. In preparing the criteria for creating management situation contracts, the corporation management group would provide a fund for investments promoting productivity advance. The means by which line managers would obtain these funds and the expected rate of return would both be part of the bidding process. What has not yet been considered, however, is who specifies the competition for the corporate management group.

Currently, mature corporations become self-perpetuating bureaucracies after the founder has left active management. In the separation of ownership and control, however, the interests of managers do not always coincide with the interests of the stockholders. Moreover, as the managers propose the slate of new officers of the corporation, the stockholder has a very difficult time expressing his concern with the management practice. Currently, poorly managed firms with hidden assets become takeover targets by financiers, who are called corporate raiders by managers. This threat of a takeover places pressure on the managers to obtain the highest return on assets. With the advance of the social nervous system, however, much more effective methods are available for stockholders as a group to exercise an ownership role in a firm.

First, to exercise their ownership role stockholders need information rights which will allow them to analyze the performance of their agents, the managers of the corporation. As a result of advances in computing, corporate performance can be analyzed down to the level of each profit center, however the corporation provides the stockholders only aggregate data in the quarterly and annual reports. To ensure that the stockholder can analyze the performance, the corporation would place in the public domain, a system of files containing disaggregated performance data maintained by the SEC[6]. This data would be transmitted from machine to machine and stored on one-time write type storage devices for public perusal. The stockholder would be able to analyze the profitability of each product line, the effectiveness of the system of corporation incentives, and the return on each type of asset held by the corporation. As was the case for the consumer, the stockholder would not exercise his right directly but rather through his evaluation service, for example, Value Line. These evaluation services would be able to analyze corporation performance in much greater detail than is currently done by either the major trading houses or by purveyors of stock market analysis letters.

Besides much better data to analyze the performance of their agents-the managers, stockholders need rights in order to exercise their ownership function. A set of stockholders rights has been proposed by the United Stockholders Association[7], a movement led by the eminent corporate raider, J. Boone Pickens. These rights include: the basic principle of one share-one vote, removal of takeover defenses such as the poison pill, stockholder access to the list of stockholders, and the secret ballot for stockholders. These rights would greatly strengthen the stockholders ability to remove inept management from office.

The exercise of a secret ballot would require a new institutional arrangement for stockholder voting. One possibility would be for the voting to be administered by an outside firm with a fiduciary trust to the stockholders. But because low cost communications greatly reduce the transactions cost of creating organizations, stockholders of many firms would develop stockholders associations to run the elections. For a well managed firm the stockholders would generally accept the recommendations of the existing officers, but in a poorly managed firm, the corporate raider of today would have an inexpensive mechanism for shifting the direction of the corporation. Competing groups for control would present their case through the stockholders' association bulletin board.

For some corporations, the stockholders would not vote for a new slate of officers but rather for the criteria describing the corporation manager's contract in the management situations market. While the existing managers would propose their criteria, the stockholders would consider their proposal one of the possible alternatives. While most stockholders, especially small stockholders, would participate indirectly in these battles through proxy, having a low cost means of communication and organization would facilitate much greater competition for setting the policy objectives of each corporation.

The market structure in which the management groups operate under the proposed design would be very different from the current structure. Upon obtaining the rights to control a production facility, the new managers would start by making the major improvements for the duration of their contract. The management group would take the environmental regulations as given and would find the best resource use of the profit center. For plants which make products for final consumer sale, the management group in many cases would be free to lease the right to produce any product in existence. In this way the management group would be determining the output mix. Using forecasts of consumer demand the manager would vary this output mix to maximize profits according to changing demand conditions. As a professional the manager would be held responsible for the conditions of the contract, including his obligations to the ownership group and to the public.

The performance measure of the management group is how well they fulfill the conditions of the contract. In seeking employment, management groups would generally consider the entire industry rather than a particular firm. Managers from declining industries would retrain in order to enter expanding industries. The management situations market, then, provides managers with total control over their careers. Moreover, managers working within such a market would know that increasingly their careers will be judged on the basis of merit.

Innovation

In developing a more flexible organization, firms will more rapidly take advantage of the opportunities created by an increasing rate of discovery, invention, and innovation. Intense international competition will force government and private firms to constantly innovate to increase this rate. First, government will develop a better allocation mechanism for research and development expenditures and gradually increase the level of funding. Second, the intuitive improvisatory strategy will gradually be replaced with a more systematic strategy for implementing innovations. Finally, public and private institutions will develop better transfer mechanisms from public research to private inventions and innovations.

The level of funding for research topics is roughly related to their perceived long-term effects in producing positive social benefits. One weakness in the current funding mechanism, however, is that with the collapse of the scientific advisors under the Reagan administration, the mechanism for establishing priorities according to a science policy created by science advisors has broken down. While improvements in the funding of science can be made by creating a better review process, funding will always have a political component. Because the future benefits cannot be accurately forecasted, no review process can precisely allocate resources based solely on marginal benefit versus marginal cost. In the next century the funding to natural sciences and engineering will gradually increase due to the pressure of intense international competition. The potential for spectacular results in the biological sciences and applied biology should markedly increase funding in these fields. And with society unable to afford medical progress, funding may shift to research leading to cheaper, more effective medical procedures.

In the past century, social evolution has transformed the process of invention from the lonely inventor using trial and error methods to an organized activity using systematic research and development methods. To a much smaller degree social evolution in the past century has transformed the process of innovation from one which depends on the heroic figure of the intuitive entrepreneur to one which proceeds as a systematic, organized activity. Given the failure of US business firms to innovate in quality control and to maintain the leadership in manufacturing, the public recognition of a need for a much more systematic approach to the problem of innovation is increasing. In the next century systematic strategies for implementing innovations will be developed to replace the intuitive improvisatory strategy.

The development of a more systematic strategy for implementing innovations requires the a much higher rate of discovery in the social sciences and applied business disciplines to develop analytic theories which can predict the consequences of alternatives. The advance of discovery which promotes innovation will be greatly improved by the observational implications of the right to learn. For scientific purposes the right to learn guarantees that an investigator is at least entitled to an unlabeled, representative sample of observations of any phenomenon. How researchers will exercise this right to balance the benefits of better observations against the possible infringements of privacy and trade secrets will be discussed in the next chapter.

The empirical study of management would improve by much more statistical testing of theories across corporations using nonexperimental statistics. An example would be a study-using an industry wide data sample, of incentive and performance data of the impact of the variation in incentives on manager performance. Representative samples might also provide data for a better understanding of consumer behavior, a topic of great interest to both producers and consumer services. As decisions become increasingly computerized, continual research is required to evaluate the decision rules. In practice, such evaluation requires a data sample of the consequences of variations in the decision rules, which generally would require a data sample from representative decision makers. While considerable research of this sort takes place today, the researchers are frequently restricted to posing questions which can be answered by the available data. Removing current data-gathering restrictions would greatly accelerate the advance of knowledge leading to innovations.

The right to observations implied in the right to learn is insufficient to guarantee progress in innovation-producing knowledge as nonexperimental statistics frequently contain too little variation to enable the research to discriminate between alternative hypothesis. To obtain greater variation systematically, the separation strategy currently used for innovation in agriculture should replace the intuitive improvisatory strategy, wherever possible. This will require building expensive research facilities to test, for instances, alternative organizational structures and incentive systems made possible by dynamic image fiber optic communications. The justification of the more expensive, systematic approach to innovation is that it would produce superior results much more rapidly.

Parallel to the current institutional arrangements for rapidly transmitting advances in the natural and biological sciences into inventions, a similar development will take place for transmitting the advances in the social and allied business disciplines into innovations. With much better observations and experimental stations, innovation research will much more rapidly discriminate between alternative theories. And with better theories, innovations will become more an exercise in applying theory and less an exercise in pure intuition. With the slow, gradual creation of bodies of theory in the social sciences, the ability to demonstrate the superiority of a systematic approach to innovation over the intuitive improvisatory strategy will in all probability increase. With the exception of controversial, political innovations, the funding for basic research promoting innovation should also increase dramatically as a systematic approach to innovation is developed.

Firms will also improve their capability to invent and innovate. One aspect of this improvement will be the elimination of the current weaknesses in American manufacturing[8]. By reorganizing to achieve much better coordination between engineering design and manufacturing US firms will achieve much better coordination between the invention of new products and the associated innovations needed to manufacture them. Also, manufacturing will attract more competent people as salaries for manufacturing executives and automation engineers are raised. Finally greater communication between the producer and customer will promote faster market driven invention and innovation.

Another avenue to invention and innovation improvement will be the sharing of risk through the creation of research and development consortia. The 1984 National Cooperative Research Act permitted the creation of research consortia between business rivals. As was discussed in Chapter One, consortia have promise in bridging the gap in incentives between basic research, with a long term economic potential, and product research and development. For basic research, which has long-range economic promise but great expense and risk, a research consortium between competing firms promotes applied research in the gap between basic research and specific product research and development. The much wider communication network than that resulting from single-firm product development will aid advance discoveries promoting invention and innovation. Research consortia will also develop for innovation in production processes which are undergoing rapid technological change. The evolution of consortia will necessarily take time as firms work out compromises between the benefits of cooperation and their inherent, underlying competition. The means by which the public interest should resolve the conflict between promoting faster transfer from research to products and promoting economic competition should be determined by observing the performance of consortia.

The advance of communication technology will aid both basic research and the transfer of basic research to the marketplace. Today, researchers involved in a particular research topic are scattered throughout universities and research centers. Because of the lengthy review process for professional journal articles, communication in professional journals is usually simply recognition for a new result. Conferences, too, are formal procedures which require a long lead time for planning and organizing. Both of these means of communication will ultimately be overshadowed by inexpensive, dynamic-image teleconferencing which will promote many informal meetings between specialists scattered at remote sites. Scientific interest groups will accordingly become communications networks.

Improved communications will also accelerate the transfer of knowledge from basic research to invention and innovation. Already universities trying to obtain a higher rate of return on research and to promote startups for economic development are creating databases summarizing university research activities. One example will be the use of image teleconferencing to promote faster transfer between basic research and applied research groups. Transfer of such knowledge to applied groups will tend to increase, because applied groups can more easily identify potential marketable ideas. Also, by promoting miniteleconferences between research groups and potential application groups, the university can reduce real cost of communication, that is travel costs and time delay.

In addition to improving communication between researchers and potential developers universities will take an active role in promoting startups by performing such services as providing help with property rights. With a high rate of advance in knowledge and an easing of the knowledge transfer problem, there will be more invention and innovation opportunities surrounding major research universities. These universities will attract consortia and high technology firms in the specialties of the university.

New inventions and innovations are created by individuals alone or in teams associated with all types of public and private institutions. The greater the change associated with an invention or innovation, the greater the probability that the invention or innovation will be created by a startup, that is a new firm. The types of changes we will be considering here are changes in production technique, organization, or customer behavior. While larger organizations will become more flexible, they are less adept at change than new organizations since skills used within an existing large firm do not have a competitive advantage in dealing with a completely new situation. Large firms can reduce the risk in invention and innovation by allowing startups to develop a new market. Firms producing completely new products, face great uncertainty concerning the performance and cost of the mass produced item. By waiting until the startups reduce the risk through learning, a large firm can enter the market later either by creating an imitation or buying out a startup. Large firms have a competitive advantage in access to capital and large scale production and marketing, hence a large firm has a better chance of competing in the race towards market share in the mass market than the initial startup.

An important factor in the dynamics of firms is the competition between firms to expand production of a successful new product in order to obtain the economies of mass production. A new market frequently starts with a large number of startups. From these, a few startups make the transition from startup to mature corporation, and in this way a new markets evolves. The transition from a startup to a major corporation is difficult, however, because it requires constantly changing organization and talents. Moreover, with economies of scale a far smaller number of firms can supply the market, creating a survival of the fittest situation. A major corporation has an incentive to enter new markets in the growth phase to promote overall corporate growth. Major corporations can compete in manufacturing and marketing. With a constant stream of new products entering the marketplace the dynamics becomes the struggle of startups to reach maturity, with older larger firms constantly repositioning themselves through acquisitions and spinoffs in growth markets.

With changes in the tax structure, startups will have an easier time in financing the transition. Startups are initially financed by venture capitalists and shift to retained earnings or equity finance as the firm prospers. As was recommended in Chapter 4, the tax burden should be shifted from income to consumption. With the end of double taxation on dividends, corporations will pass more profits on to stockholders and retain less as retained earnings. Thus, individual investors will make more choices in picking the potential winners among startups. With the move to deregulation, investment banking should return as means of financing the transition.

Dynamics: Firms and Work

Considering the dynamics of the marketplace at a single point in time, the number of firms of a particular size will bear an inverse relationship to the size of the firms. Service activities are likely to see the evolution of a greater number of large firms. One trend will be for the evolution of more large nationwide, or even international, partnership firms for professional services. The contemporary legal profession is making this transition. A second trend will be the evolution of more national corporations for services. This trend is evident in the provision of medical services. For example, with the trend towards prepaid preventative medicine, insurance and other corporations have created health maintenance organizations, HMOs. Also, large corporations will evolve to create the numerous evaluation and producer services discussed in the text. For example as the digital capacity of the communication system expands information utilities will become large corporations.

In spite of the fact that large firms will become more flexible, the role of small and middle sized firms will increase rather than decrease. To achieve flexibility, large firms subcontract, creating greater roles for smaller firms in the marketplace. With a more analytical approach to considering the alternatives in the marketplace, a smaller firm with a superior product will have a better opportunity than in the current market situation. The right to learn will also create more opportunities by creating niche markets. Using the right to learn small firms have the right to open interfaces, which means is that if, for instance, IBM creates a new computer, it must release the specifications to outside software concerns at the same time that it provides these specifications to internal software groups. This would enable firms that have carved out a niche in software to compete on equal terms with IBM. As software grows in importance throughout all aspects of business, the creation of industry standards, for example communication standards for local area networks opens niches for smaller firms. With an open interface policy, smaller firms have opportunities to provide specialized products to smaller markets.

The great difference between the development of products today and in the future will be the amount of investment in computer programming. The new factor in assemblyline production is the control program which runs the robots and coordinates their activities. Programing tasks will similarly create services, which will increasingly take the form of user friendly programs. This aspect of product development greatly changes the method of productivity advances in factory production or services. As a result, one important aspect of improving productivity will be improving the programs. Even in factories employing no people, better programs facilitate more efficient machine use. In improving many service programs, the aim will be to provide with more features or to make the code more efficient. Modifications to products, moreover, will come about by modifying the control programs. With an open interface policy, these objectives can be met by a dynamic market for add-on software either to improve the production process or to add new desirable features to current products. With an open interface policy where a software developer knows the interfaces and the potential market by observing alternative production sites, he can decide the potential profitability of various add-ons.

As has already been explained, most people in informational society will be employed in smaller firms[9]. With the advance of automation, routine work will be displaced first, while work and conditions of work will be increasingly varied. With constant change, jobs will tend to exist for fixed time periods such as the duration of the firm's contract to supply a product or service. Startups which are expanding to become major corporations can guarantee long term employment, but once the new industry matures, automation will decrease the demand for workers forcing the former startup to devise a strategy to retrench. Increasingly, then, workers cannot count on lifetime employment in a particular industry. When employment in an industry peaks, workers must start retraining for the next step in their career within a new expanding industry. Under these conditions the life of the worker, whether blue-collar, information, or professional, will tend to cycle between work and study to prepare for the next job. Workers who, like managers, work long hours while on the job, will tend to cycle between full-time study and full-time work. In work situations where the worker works a short work week, he will very likely pursue both activities simultaneously.

The social nervous system provides the future worker with two advantages over today's worker. First, much of the training to prepare for a new job can be accomplished at any terminal through the social nervous system. A worker, as a student, could obtain courses from practical instruction in how to operate new software to the most advanced post graduate instruction. Only for training sessions actually requiring special type equipment would workers have to travel to a specific location; however, with advances in virtual reality workers would only have to travel to a specific location for the final stages of such training. Second, to achieve low cost matching of supply and demand in fluid labor markets, the matching will take place through the social nervous system. Both the available workers and the available jobs will both be in databases addressable by any terminal. In industries where performance standards have been established the file on potential workers will contain sufficient information to apply analytic criteria to quickly screen potential applicants. The time required to find a new job will be greatly shortened.

As workers gain greater freedom of location they can search for jobs over a much greater physical area without relocating their home. As the work week shortens, a trend in work patterns for physical work will be to work concentrated periods and then have longer periods with no work. Such workers, for example, might work one week with 10-hour days and then have two weeks off. As physical plants in an industry frequently locate relatively close together, they will tend to create orderly markets for physical object workers, as these individuals can work at any plant and live a considerable distance away during their off periods. For workers manipulating informational objects in the social nervous system, freedom of location would be even greater. The counterpart to the current salesperson in a department store is the information worker who would come on screen to help people who were unable to use a particular aspect of the social nervous system software. Such workers would tend to work during the peak demand periods. With cheap communication, the change in time zones can be used to smooth out employment hours. And as communication costs fall the potential pool of information workers increases.

With a more efficient information system to match supply and demand and with a greater pool of potential workers for every job, market forces would quickly clear the labor markets. In response to the fragmentation of work, the ability of unions to set wages will decrease markedly, and the market itself will replace union contracts. This does not mean that unions would disappear totally, however. The role of unions and professional societies in markets would shift to ensuring that the market clearing processes satisfied due process and the workplace satisfied the safety code. The union would also act as a special interest group for workers' legislation.

Provided that society maintains a high level of investment, average income over lifetimes should increase; however, with short-term work and decreasing work hours, work for all economic workers whether owners, managers, professionals, information workers, or blue-collar labor, would be much more professional and much less personable and would offer very little day-to-day job security. To maintain a stable level of consumption with fluctuating incomes all workers would need to be knowledgeable about finance. However, the more personable lifestyle in the new towns would compensate for the less personable nature of economic relations.

And, with a decreasing work week most workers will be more oriented towards their communities and less towards the workplace. Moreover, with shorter, more flexible work hours and part time telecommuting to work and school, working couples with children would find it much easier to organize their schedules to promote family life than currently.

Economic Performance

Having surveyed the possible changes in the economy of informational society, we will next consider some aspects of economic performance. In comparison with current society informational society is likely to have an even faster rate of discovery, invention, and innovation. Expenditures on basic research are likely to increase and with improvements in the transfer mechanism from basic knowledge to invention the flow of new products and processes should increase. With a more systematic approach to innovation and the tremendous opportunities in automation and communications the increase in the rate of innovation should be even faster.

The increase in the rate of discovery, invention, and innovation is likely to increase the randomness of economic behavior. An advance in basic knowledge has a large multiplier in practical knowledge as a scientific advance is transformed into social advance through application. The less capable theory is at explaining behavior, the more of this knowledge must be gained by trial and error, learning by doing, and imitation. While the social sciences and applied business disciplines will advance somewhat, improving the ability to predict the consequences of alternatives without experimentation, the overall improvement will not be enough to compensate for the increased rate of discovery and invention. This experimentation required to improve performance is generally performed by risk takers such as trend setting consumers and startups in production. In activities where theory can not accurately predict behavior, the variation in perceptions among the risk takers will produce the appearance of random behavior. The economy of informational society will have much more variation in economic behavior than the current economy does, as risk takers search for better alternatives.

A requirement of perfect competition is that all the economic agents have perfect knowledge of their alternatives. With an efficient set of information rights, all the economic agents would have the potential for much better knowledge of their alternatives than they currently do.

What will realize this potential is the growth of evaluation services to provide the various economic agents data with low-cost analytic evaluation of alternatives. The combined forces of a growing the social nervous system, efficient information rights, increasing analysis, and the development of evaluation services should move the market to a position much closer to the information requirements of perfect competition in which each agent has perfect knowledge of all his alternatives.

Moreover, the improvements in decision-making afforded by analytical comparison of alternatives discussed in this chapter will make the behavior of bounded rational decision makers more nearly rational. Information services, especially the quality ones, will provide their clients with decision- making aids and organize the data in reduce the limitations of their clients cognitive skills. These improvements will greatly improve the ability to decision makers to analyze large numbers of alternatives. With unaided intuition the bounded rational decision maker, when faced with a decision task involving a large number of alternatives, frequently resorts to superficial criteria in order to pare the number of alternatives to a more cognitively manageable number. With analytic comparisons made possible by cheap computation, however, the decision maker can screen the entire sample of choices with whatever screening function he wishes. With producer services offering criteria for alternative decisions, the decision maker can interactively examine the consequences of many alternatives, thereby improving the quality of decision making, since he would have a greater probability of finding a better choice among the possibilities which he would otherwise have superficially dismissed. The decision maker's ability to identify all his alternatives and to accurately assess their properties depends, in turn, on his information rights. Given the information rights suggested in this chapter, private entrepreneurs will be able to create databases to support most common decision making.

A consequence of a more rational consumer with appropriate information rights is that the economy will move closer to the theoretical ideal of consumer sovereignty, which means that markets are driven by consumers' choices. In current society consumers' choices are influenced by advertising thus the system is closed. In the projected society, however, the consumer will have low-cost information services to find the best price and through his information service the consumer can analytically analyze the qualities of alternative products. This means that the role of analysis will increase and the role of advertising based on pure emotion will decrease. This is not to say that advertising to promote new products will cease, but simply that argument based on pure hype will be less effective. Instead the information structure will quickly locate good products and reward them with high sales. Poorer quality products, on the other hand, will have to lower their prices or fade from contention in sales. Within such a structure, the consumer's role would more closely approximate the theoretical role of consumer sovereignty.

Adjustment to changes in supply and demand conditions should be much more rapid than is currently the case. With most market-clearing operations taking place in the social nervous system, price adjustments to changes in demand and supply will be almost instantaneous. Furthermore, as many services move to the social nervous system, the market for such services will become international as the social nervous system itself acquires international dimensions. This means that any local, regional, or even national variation in demand can be met without relocating the supply. Also, with multiproduct plants, the output can be shifted by changing the control program. Over time the range of products producible at a single plant should increase. This means that output can shift more quickly than currently to meet changes in demand conditions. For example, eventually automobile manufacturers could quickly shift production from small to large cars if demand conditions changed.

Not only will output shift quickly with changing demand conditions, organizational structures will also adjust much more quickly to accommodate change. The information policy creates the framework for a much more dynamic corporation culture than currently exists. With better information, stock market analysts would identify poorly managed corporations much more quickly than they can at present. And as information policy facilitates takeovers, the corporate management team would be under much greater pressure than today to obtain the highest return on equity. In response to this pressure, the management situations market for line management would facilitate the constant re-evaluation of production facilities and software.

Notes and References 6

1. The market is moving towards providing more background information on products in the evolution of computer shopping services. Prodigy, a second generation computer shopping service, provides the user with the information from the Consumer Union testing service. Shopping services such as Prodigy have the same inherent conflict of interest as the airline reservation systems. Sears is one of the corporations behind Prodigy and since Prodigy will offer

many stores Sears interests in selling its own goods are in conflict with efficient operation of the information service.

2. Crowe, Michael, 1988, *A General Study of Corporate Flexibility*, Senior thesis, Department of Economics, The University of Texas at Austin

3. One of the early contributions to whether an activity would be performed inside a firm or through a market is Coase, R.H., 1937, The nature of the firm, *Economic N. S.*, Vol 4, pp 386-405. Coase's ideas have been amplified by Williamson in his development of transaction-cost economics. For example, see Williamson, O. E., Vertical Integration and Related Variations on a Transaction-Cost Economics Theme, 1985, in Stiglitz, J.E. and G. F. Mathewson (eds), *New Developments in the Analysis of Market Structure*, (The MIT press, Cambridge). For Williamson and others an important explanation for the existence of the firm is asset specificity. In informational society asset specificity decreases for two reasons. First, more economic activities take place through the social nervous system independent of location and second, soft automation makes fixed assets more flexible. This means that more activities will take place through markets.

4. Staff, 1986, The Hollow Corporation, *Business Week*, Mar 3, pp57-85

5. For a discussion of the intricacies of internal resource allocation within institutions see: Hoenack, S. A., 1983, *Economic Behavior within Organizations*, (Cambridge University Press: Cambridge) The technological advances of informational society reduce the information costs of the participants in organizations.

6. The problem of deciding exactly what data should be released is a problem in information policy, which will be discussed in the next chapter. As Stigler points out required disclosure by the SEC has had a questionable impact on investor performance. See: Stigler, G.J., 1964, Public Regulation of the Securities Market, *Journal of Business*, 37, pp 117-142. In view of the subsequent development of efficient market theory, this finding is not surprising. Subsequent research indicates that SEC required disclosures are more successful at reducing risk. See discussion in Posner, R.A. and K.E. Scott, 1980, *Economics of Corporation Law and Securities Regulation*, (Little, Brown and Company: Boston). In informational society the cost of providing data by machine talking to machine will fall greatly. The objectives of information policy to investors are first, to reduce volatility by increasing the accuracy of analysts short term earnings forecasts. And second, to reduce the bias for short term performance by providing analysts with better information to judge long term performance.

7. Their views plus many others are contained in: Subcommittee on Securities, 1992, *Shareholder rights : hearing before the Subcommittee on Securities of the Committee on Banking, Housing, and Urban Affairs, United States Senate, One Hundred Second Congress, first session, on the responsibilities of the board of directors in protecting the rights of shareholders, the proper role for shareholders in corporate decisionmaking, the best corporate governance structure to endure long-term growth, and the disclosure of executive pay, October 17, 1991*, (U.S. G.P.O.: Washington)

8. A complete list of improvements is to be found in Dertouzos, M. L., R.K. Lester and R.M. Solow, 1989, *Made in America: Regaining the Productive Edge*, (The MIT Press: Cambridge)

9. The assumption is the continuation of current trends. For example, see Birch, David L., 1987, *Job Creation in America: How Our Smallest Companies Put the Most People to Work*, (Free Press: New York)

Chapter 7

Information Policy

Introduction

For the next hundred years liberals and conservatives will engage in numerous political disputes over the proper formulation of information policy. A recent example was the dispute over requiring firms to give workers sixty days notice before closing a plant[1]. Numerous factors will intensify the conflict. First, in an increasingly complex interdependent political economy, the information desired for decision making is frequently held by others. Second, in a highly regulated political economy, observations must be made to determine the behavior of the political economy in order to more rationally regulate the behavior. Finally, the rapid advance of information-processing technology is creating new conflicts between privacy and efficiency not addressed by common law or statute. As information processing technology advances, resolutions of the conflicts between efficiency, science, privacy, and property rights must be constantly reconsidered. Because most groups in society are affected by the resolutions, information policy is and will become even more contentious.

To consider information policy, envision the decision maker seated at his terminal which is connected in a huge network to the computers of all the other decision makers. As the quality of many of his decisions would be improved by access to information held in files by other decision makers, the relevant question is what information should he be able to access? This question is the fundamental problem of information policy. Operationally information policy should promote efficient markets, effective government, and accountability of public and private officials. In addition to these operational goals, a scientific goal of information policy is to increase rate of discovery in order to promote innovation.

Achieving the operational and scientific goals of information policy is in direct conflict with other social goals such as privacy and property rights. Increased economic efficiency, for instance, depends on third parties creating databases containing detailed information that allows the decision maker to predict the performance of alternatives. But in facing the prospect of third parties with vast databases containing detailed personal information, the individual at the very least feels a loss of control. Similarly, firms fear a loss of trade secrets. Consequently for many individuals and groups the value of economic efficiency and government effectiveness pales in comparison to the potential dangers of information abuse. Information policy must therefore balance the contradictory economic and political goals within the alternatives made feasible by information technology.

To formulate an appropriate information policy let us start by quickly reviewing the development of information policy up to the present. The one regular data-gathering function of the executive that has been explicitly stipulated in the Constitution is the census every ten years. For the purposes of formulating law, the legislature has the power to obtain information through investigations. This power was not made explicit in the Constitution, but was implicitly assumed in keeping with parliamentary traditions established in the 17th century[2].

The primary concern in information policy at the end of the 18th century as expressed in the Bill of Rights was to prevent possible abuses of government power. The First guarantees free speech, which restricts prior restraint. The Fourth Amendment restricts the governments to search and seize persons, their houses, papers (information), and effects and the Fifth prevents self incrimination. The Fourth Amendment establishes restrictions in obtaining information in criminal investigations. Law enforcement agents' need a warrant based on probably cause to conduct a search. An individual can use the Fifth Amendment to deny the legislature information which might incriminate him, unless the legislature grants the individual immunity from prosecution.

The constitution in prohibiting the states from interfering with contracts greatly strengthened property rights, which includes informational property rights. These rights are the bases for many current services that sell information to aid in decision making. Two of the oldest such services are stock quotes and credit information. A more recent example is airline reservation systems. The Constitution further promoted information creation in authorizing a federal system of intellectual property, patents and copyrights.

Privacy was a private not a public concern. For example, as there were no models for using data from market transactions and as transactional data was costly both to acquire and to process, there was no need for a specific Constitutional provision for privacy. Restrictions on the release of transactional information were regulated by market forces such as the general acknowledgment that confidentiality was a desirable quality of a transaction.

The growth of information policy since 1789 has been an *ad hoc* reaction to events. One example has been the evolution of administrative agencies' power to obtain information through investigations, required reports and inspections. With the creation of regulatory agencies, starting with the ICC in 1887, the legislature granted these agencies broad powers to investigate within the mandate of their authority. The investigatory power of regulatory agencies such as the Federal Trade Commission was initially checked in the 1920s by court interpretations, but subsequent court decisions reversed these decisions[3]. In current society administrative agencies have, in order to set administrative policy within their mandates, the same powers to investigate as the legislature. To ensure compliance with the law, administrative agencies have broad powers to impose required reports. For administrative activities such as the enforcement of building codes and health and safety regulations, government agencies have in many cases obtained the power of inspections without search warrants. Inspections without warrants is not universal as OSHA must obtain a warrant to conduct a safety inspection[4].

Another aspect of information policy is required disclosure by firms. One purpose of required disclosure is to provide individuals information to improve decision making. A secondary purpose is to regulate the behavior of firms through public awareness. Food products, for instance, must display the components ingredient; money lenders must provide the simple interest rate[5]; and corporations must reveal important financial information to stockholders. A recent addition to disclosure policy is the law granting workers the right to know what hazards exist in the workplace[6]. This right to know has been extended to the communities right to know what pollutants are being released into the community by businesses. In some cases such as gas mileage data the government generates the data and then requires that automobile dealerships display the data on new automobiles for sale.

Besides obtaining information passively through required disclosures, individuals can actively seek information using information rights to access files on request. For the purpose of making government more accountable, the Freedom of Information Act[7] gives individuals broad powers to access federal government records. Many states have enacted similar legislation. Also, shareholders have the limited rights to access corporation books[8]. These later rights are defined by the state in which the corporation is incorporated.

As technological advances have created the capacity to impinge on personal privacy steps have been taken reduce the abuse. One example is the tort for invasion of privacy in response to the excesses of sensational newspaper reporting in the late 19th century. This common law approach to privacy creates the right to be left alone. The tort has four elements[9]. First is an intrusion of privacy. Second is the appropriation of ones likeness for commercial purposes without ones permission. Third is public revealing of private facts and four is portraying one in a false light.

Additional concerns for privacy have arisen with the creation of numerous databases which has been greatly promoted by the increasing digitization of records in all institutions. The low cost of transferring and storing information in this form makes feasible the creation of data bases containing detailed information about behavioral characteristics. These extensive datafiles pose serious problems for abuse and intimidation.

To cope with the privacy problems of databases, information policy has evolved further safeguards[10]. The federal Privacy Act provides individuals the right to inspect and correct personal information in federal datafiles. Many states have enacted similar legislation. The Fair Credit Reporting Act gives individuals the right to inspect and correct their credit files maintained by firms. Individuals do not have, however, a general right to inspect and correct personal data in privately held datafiles.

Another safeguard for information in datafiles is a specified release policy. To protect firms' trade secrets commercial or financial information is exempt from the Freedom of Information Act. Most business data released by government is aggregated to the industry level to mask the behavior of individual firms. Medical doctors, with some exceptions which vary from state to state, do not have to reveal medical record of their patients. The release of school records is regulated by many states.

A final aspect of information policy is a "can not use" category for certain information. Two examples of information classified in this way are insiders information in stock trades and prohibition on the use of race, creed, sex, age and other variables in employment decisions. In databases this "can not use" criterion means that certain data must be purged from the database. Much negative credit information must be purged after seven years. In Maryland an employment file can not contain psychological information not related to an employee's capacity to perform the job.

The growth of information policy, then, has been a piecemeal response to particular political problems. Information policy, today, is an *ad hoc* collection of property rights, required disclosures, information rights, and restrictions. Improving political economic performance is strongly dependent on having an information policy appropriate for the information technology. To design such a policy, the goals and alternatives of information policy need to be analyzed from a general framework.

Economics of information

With advances in information technology, the most important aspect of information policy will become the policy for decision-support systems. Increasingly decision support systems will be used in all types of decisions whether personal, economic, or political. The problem to resolve is what information should the decision maker, seated at his terminal, be allowed to access, given the premise that he would be technologically capable of accessing all information. This problem has an operational and a scientific component. In ranking alternatives the decision maker uses criteria to rank alternatives. The operational question is whether he can obtain the information in order to use the criteria to rank the alternatives. The scientific question is whether he can obtain the data to evaluate and improve the criteria.

The starting point to resolve this issue is the question of how well will economic incentives alone achieve the operational and scientific goals. Decision makers demand information to rank new alternatives created by constantly advancing technology. Such information, however, is often costly to acquire and to process. Accordingly, to rank his alternatives each decision maker must consider the costs and benefits of acquiring and processing information. From experience he learns to forecast the benefits and costs of obtaining alternative information and as a bounded rational man he employs rules of thumb to obtain the correct amount of information for a decision. The cost of collecting and processing information have been dramatically reduced by the advances in information and communication technology; hence, most decision makers in informational society will collect and process much more information than prior to the introduction of information technology.

Many decisions have common features, a fact which creates great economies of scale in collecting and processing information. This situation provides an economic incentive for entrepreneurs to create databases to which they can sell access to interested decision makers. Given the scope of the potential markets, entrepreneurs will tend to specialize in particular types of common problems. While the decision makers using rules of thumb may err toward collecting too much information, to the extent that his behavior approximates rational man, he has no interest in collecting extraneous information. The entrepreneur, then, in creating a database has an incentive to collect only information considered valuable by potential clients.

For comparing large numbers of alternatives the amount of information considered valuable depends on the decision technology employed. Using unaided intuition, a decision maker would not want to process more than a few variables on each subject. In evaluating alternatives analytically the demand for information depends on the criteria employed for numerical processing. To analytically search to find the lowest price requires a single variable, whereas to employ an expert system to evaluate loan applications requires numerous variables. As more complex criteria are demonstrated to be more effective in evaluating alternatives than unaided intuition and single attribute evaluations, the demand for information will increase.

Economic incentives will induce database services to focus increasingly on selling value-added services as opposed to raw data. Database owners may have property rights to the information they have created; however, these rights are difficult to enforce, especially with regards to the giving away of information free to friends and associates. Because information can be reproduced and disseminated at very low cost, database services need to provide more than simply raw data to ensure a return. For this reason database entrepreneurs will focus on creating value-added services such as criteria software to evaluate alternatives and, as new decision rules and expert systems are created to analyze alternatives, database managers will offer them as software packages in order to analyze their databases. Such entrepreneurs will make most of their profits from software processing accessed data.

Before databases can promote economic efficiency, database managers must be able to obtain the requisite information. Consider the consequences of granting each person–that is, each legal entity–an absolute right to the release of his own information. Such a policy has a positive bias since persons will provide at their expense information which reflects positively on their interests and would restrict access to that information which reflects negatively. To be sure, persons do have incentives to make a public relations release of negative information which will become public knowledge in order to minimize the adverse publicity. But in general, negative information will be more costly to acquire than positive.

For example, competition in advertising does provide the consumer with some useful information concerning alternative market choices, but market competition will seldom reveal negative information concerning products. A pharmaceutical company whose drug has a negative side effect in 1 out of 100 cases has no incentive to advertise superiority over a drug which has the same negative side effect in 1 out of 10 cases. The negative publicity resulting from such advertisement would probably decrease the demand for both products and might prompt action from the Food and Drug Administration. An example of a public relations release of negative information would be a politician who, knowing he could not keep the fact that he once smoked a joint of marijuana secret, admits this fact as a mistake of his youth.

The objective of operational information policy is to incorporate both positive and negative externalities into decision making. In repeating-type decisions, such as purchasing a product or performing repeating tasks, negative information will be revealed and will become public information over time. Thus, if technology were static over time, both positive and negative externalities would be incorporated into decision making. The more rapid the rate of change in technology, however, the less time negative information has to accumulate before the next change. Consequently, the amount of bias is proportional to the rate of change. Economic incentives alone are insufficient to ensure that decision makers can acquire negative information to rank alternatives.

Similarly, economic incentives alone are unlikely to ensure that decision makers can obtain the information to evaluate and improve criteria. Consider, for instances, an expert system for evaluating loan applications. Ideally an investigator would like to conduct a controlled experiment comparing the performance of alternative expert systems with loan officers of various intelligence, training, and experience. Baring this, the investigator would like to obtain a representative sample of the past performance of the various alternatives. To use past data the investigator would have to predict the consequences of how a loan applicant who was actually denied credit, would perform if he were granted credit under a different evaluation procedure.

Scientific data and operational data have very different characteristics. For operational purposes the decision maker needs to identify the individual and needs the specific information required to apply the criteria. For scientific purposes the investigator needs only a representative sample and has no need to identify the individuals. For many scientific purposes the investigator, however, would like a large number of variables to test alternative hypotheses. As long as individuals fear that scientific information might be used in some operational capacity against their interests, they will have little incentive to provide potentially damaging information. Firms will never have an incentive to provide information which might benefit competitors. Economic incentives alone are unlikely to produce representative data samples.

To achieve the objectives of information policy decision makers and scientists need information rights to obtain information, and the basis for these information rights will be the creation of a right to learn for all legal entities. Specific operational and scientific information rights will be derived from this general right to learn through legal precedent and legislative statute. For operational decisions the right to learn provides the basis for creating an information policy which enables decision makers to analyze alternatives. To promote innovation the right to learn implies an information policy which provides observations on all phenomena. This observation right is the basis for empirical science in the political economy.

Information rights granted under the right to learn must, however, be balanced against other social concerns. If information rights were absolute, there would be no trade secrets, private decisions, or national security concerning the military. Taken absolutely, an observation right would give an individual the right to observe his neighbor's bedroom behavior at will. Obviously, compromises must be made between information rights and other social concerns, such as property rights, national security and personal privacy[11].

Operational information policy

The right to learn provides the framework for creating operational information policy governing decisions based on current knowledge. Operational information policy should promote market efficiency, government effectiveness and the ability of principals to evaluate agents. With regard to this third property of operational information policy, we might cite the examples of stockholders who must evaluate management and citizens who must evaluate government officials. Most of the types of decisions under consideration will in informational society be addressed by analysis using a decision support system. As the social sciences, the business disciplines, and artificial intelligence advance, they will create an increasing number of models for analysis of alternatives. In this context the issue becomes the extent to which information rights should be created to provide input for models to evaluate performance.

Consider first the problem of information policy as regards the promotion of economic efficiency. Advancing information technology will greatly increase the information available to decision makers. To appreciate this point, we need only consider the consumer who evaluates alternative automobiles. As the business records of repair shops are computerized, and as more sensors are built into automobiles to determine the current operating condition, the buyer could obtain a complete repair history and current sensor status report. The buyer of a new automobile could potentially obtain all test reports in the automobile manufacturer's research and development efforts at very low cost. Whatever information is made available to the buyer as an information right, consumer services would use to predict performance through the creation of models.

Consider also the problem of an employer evaluating prospective employees. As records in firms become more detailed, the creation of analytical measures of performance will become commonplace. The more data of an applicant's past performance made available to prospective employers the better models can predict future performance. As information processing capacity expands for businesses and individuals, market decisions of all kinds will change. For comparing large numbers of alternatives in the marketplace, decision makers will use analytic models or tools at least for an initial screening to obtain a small number of candidates which can then be subjected to a more comprehensive, intuitive evaluation. For market efficiency, decision makers should have information rights to the input of models that have demonstrated significant prediction performance.

Given the appropriate information rights, how are the various parties going to acquire the requisite information? As the informational aspects of markets move to the social nervous system, the analysis of alternatives will take place through databases. Given information rights to data for the purpose of prediction, these markets will be much more efficient than current markets are. To compete with one another, producers will need to have their products listed in the databases of consumer services, and workers will need to have their resume information listed in job search databases. Once the guidelines for appropriate data are established, parties will voluntarily release this data, and only in very exceptional cases would the courts be used to obtain data.

As markets are currently organized consumers do not have the resources to take a producer to court to obtain the test results of a new product, nor are the benefits of this action to the individual consumer worth the cost to the individual consumer. If, as predicted, consumer services become a major factor in informational society, they will have both the resources and the incentive to obtain all information granted under information rights. In labor markets, on the other hand, the penalty of not being considered for employment is so severe that it is very unlikely that the courts would have to be used to obtain information.

Overall, then, with properly designed incentives and penalties imposed for failing to comply with information policy, all involved parties would have an interest in voluntarily complying with the information rights. Accordingly, the information requirement for a valid contract would go beyond an absence of misrepresentation to include compliance with information rights. In the course of time precedent and statutes would have to establish liabilities for failure to inform.

The effectiveness of information policy depends on the technology for evaluating alternatives. For example, the effectiveness of an information right to know the true simple interest rate on a loan to promote comparison shopping depends on the cost of obtaining and ranking alternative true interest rates. Such a policy will be much more effective when the customer can use a computer to find and rank the alternatives than when the customer has to obtain the information by making personal trips to each bank. Similarly, the information right to unit prices in grocery stores to promote comparison shopping is of dubious value in that consumers must still expend considerable mental resources to make the comparisons. Once grocery shopping shifts to the social nervous system, price comparisons become much easier to make. Providing consumers with the research and development reports of new products, today, would be costly and might not promote a better comparison of the alternatives as most consumers could not understand the reports.

As the ability to analyze alternatives by computer increases, government can reduce regulation by expanding the flow of information. As consumer services grow in informational society their ability to digest information and present it to their clients greatly changes the effectiveness of information policy. For example, a low cost policy to reduce defects in products would be to require producers to disclose their safety and other test results. Consumer services would interpret them for their clients, and manufacturers would be much more reluctant to release new products until known defects had been corrected. The demand for a direct government role in product safety would accordingly be reduced.

As the information aspects of markets move into the social nervous system the form of information requirements should be modified to improve effectiveness. Consider the labeling requirement that manufacturers list the ingredients of food on the containers. When the purchase of food shifts to the social nervous system, this type of information should become a datafile in the social nervous system. Consumers with particular diet requirements could then incorporate the diet requirements in their market basket search procedures. This approach would be less costly to the manufacturers and would be of more use to the consumer. Since the manufacturer would only need to list the components of his product once instead of posting them on every package, the listing could provide even more detailed information such as a complete chemical breakdown in addition to the food components.

Likewise the current trend to provide information of hazards in the workplace should provide a complete datafile which can be analyzed by the worker's union or professional society. An information policy to make externalities public, will cause private parties to incorporate the consequences of externalities into decision making. Creators of negative externalities would have incentives to reduce their effects. Because aircraft disasters and nuclear power failures can cause such massive damage, such a policy would not totally eliminate the need for government regulation.

Information rights can also make markets more competitive by creating an open-interface policy in hardware and software. As manufacturing proceeds towards integrated automatic systems, the ability to compete in the market requires knowledge of the interfaces linking the various equipment together. A policy of open-interface creates market niches for smaller companies to develop specializations. With an open-interface a small company can be assured that a specialized product will work with other equipment. An example of this type of market are the numerous small companies making specialized boards for personal computers. Similarly, the interfaces in software products should be open. This creates market niches for add-ons in both manufacturing and the services.

In a competitive market, the profit signal adjusts supply and demand. Suppliers, in adjusting supply, will move from markets with lower-than-average profits to markets with higher-than-average profits. Information policy could therefore be used to promote a more rapid adjustment of supply and demand by making this information-that is, the profits of a firm by product-public information.

Besides making markets more efficient, the advance of the social nervous system creates the conditions for a more efficient regulatory apparatus. The first step in constructing a more efficient government regulatory apparatus is to organize the entire system of government laws, regulations and incentives in the social nervous system, such that any private actor through his terminal can quickly determine what laws are applicable to any proposed private act. In order for these conditions to exist, of course, the laws and regulations should be written in language that is understandable to the interested citizen. Private information services will create user friendly software to analyze alternatives.

Furthermore, bureaucrats will increasingly set rules based on analysis rather than exercise personal discretion based on intuition. Although these rules might be quite complicated, for instances, taking the form of an expert system, the private actor could nevertheless readily determine his options because they would be clearly communicated to him. Provided he follows the instructions, then, the private agent is free to act. This means that the private citizen can calculate the impact of the government business interface and internalize it in his own decision making. Not all cases can, however, be covered by clearly defined regulatory procedures. New situations will arise which will require careful study before rules defining permissible behavior can be constructed. In the interim, bureaucracies will continue to use discretion. As new knowledge is acquired, the regulatory process should change in an orderly fashion. Because the ability to acquire new knowledge will require observations on business processes, business will face a large increase in demands for information necessary to study the government-business interface.

This demand for information in addition to the demands for information for regulatory purposes might seem to impose an even greater paperwork burden on business than currently, but several factors will, in fact, mitigate this trend. First, most information flows from business to government will take the form of machine talking to machine. Second, with software creating the required information, the paperwork burden will be automated.

The amount of information that should be available through disclosure and access in order to maintain accountability of private and public institutions will change with advancing knowledge. In general, the information right for accountability should enable judgment of decision-making performance. For the stockholder, this would require a freedom of information act similar to the ones which have led to the opening up of government records, since the concept of freedom of information in government lays the foundation for public accountability. By accountability we mean much more than being able to discern whether agents are obeying the law. In a private corporation the stockholders should be able to judge how the firm is performing both with respect to long-term as well as short-term profits.

Protection from abuse

With the growth of databases the potential for the abuse of information will grow correspondingly. Markets for information will tend to be self-regulating, in the sense that the database manager has no incentive to collect information not pertinent to his clients' decisions; nevertheless, potential problems need to be considered. First is the problem of creating incentives for accuracy in databases. Second is the issue of how much privacy in the sense of control over the release of information individuals should have, given the vast amount of data concerning individuals which can be stored in databases. Third is balancing the needs of secrecy such as privacy, trade secrets or national security against the benefits of better operational decisions.

Accuracy of information in databases regulated purely by market incentives depends on the value of accuracy to the clients and the cost of obtaining accuracy by the database firm. The interests of accuracy, however, of the database manager, the clients and the subjects are not identical. The database manager will use rules of thumb to equate the cost of obtaining accurate data to the value to the customers. If the accuracy of the database is sufficient to yield better decisions than other alternative approaches then the database has value to the client.

Consider, for example, a database for making credit-related decisions. If a credit database is 99.9% accurate but tends to err in erroneously denying credit, such a database would be useful for credit allocation, since to credit issuers the loss from a bad loan is greater than the loss of foregoing a small number of potential customers. Given an efficient estimator of credit and a large number of subjects, the market would be almost efficient except that the interests of the .1% would be damaged in that they would be denied credit. A low cost procedure to protect the interest of that 0.1% is to grant subjects information rights to correct their files and to establish procedures for resolving disputes.

This approach, in fact, has already been taken in credit files, but the question remains, to what extent should the database manager be liable in a tort action? This question, of course, is difficult to resolve. With no liability the database manager has no interest in increasing the accuracy of subjects' data beyond the accuracy desired by his clients. In fact, because subjects have strong incentives to correct negative information, no liability creates an incentive for the database manager to transfer the problem of accuracy to the subjects. Making the database firm liable for the expected damage of inaccurate data raises the cost of collecting data and consequently the cost of obtaining credit.

The right to learn itself implies no limit to the information database managers could collect about their subjects. As has been pointed out the natural limit is determined by a cost versus value decision. With information rights implying a disclosure of negative information the amount of information to be gathered could become very large. For inanimate objects the prospects of very large amounts of information in databases is not personally threatening, however for humans the collection of this data can lead to possible mental anguish if negative information should be widely disseminated. This aspect of privacy has a common law protection in tort, however, this protection, is directed at the inappropriate dissemination of information rather than the collection and use in decision making. The important issue which needs to be addressed is to what extent should subjects of databases have a right of privacy in the sense of prohibiting the collection and use of personal data.

The proposed compromise between privacy in the sense of restricting the collection of personal data and economic efficiency is that individuals should be protected from the collection and use of extraneous information in databases. The development of implementable criteria to limit the collection of personal data must be based on how decisions are made using databases. Such decisions generally involve an analytic screen using an analytic criteria followed by a second more intensive intuitive evaluation. In making the analytic screen the decision maker must incorporate into his criteria such social concerns as no discrimination based on race, sex, or creed. For analytic decision making the information rights in operational decisions stop at inputs which current knowledge has demonstrated to be relevant. For this purpose relevance is not subjective opinion, but rather relevance is demonstrated by a professionally recognized study.

As knowledge advances, moreover, the requirements for demonstrating relevance will advance. In the future, then, the minimum requirement might be a statistical study based on a nonexperimental design published in a journal subject to a professional peer group review. This possibility implies that while relevance might not always imply causality, it, at least, implies correlation. Data which does not meet this standard is extraneous and should not be collected in databases for operational decision-making.

The creation of databases to assist in the search for new executives or other types of employees illustrates the conflict between obtaining information to create efficient markets, human desires for privacy, and firms' concerns about trade secrets. Efficient job market databases for searching for employees analytically must contain measures of past performance that can be used to predict future performance. This criterion would require that performance information from firms be released to databases. In addition, databases would need industry standards for making performance measures or at least for comparing performance measures of alternative firms. Firms would object to the release of such data on the grounds that other firms would hire away the firm's best employees.[12]

From the perspective of market economics, social justice is an efficient market in that each worker will receive his highest compensation. In a job market situation where employment typically lasts for short time periods, the information requirements for creating efficient markets should override any misguided notion of protecting corporations from competition in labor markets. Each firm would occupy the same competitive position, and compensation would better reflect merit.

But with the emphasis on sufficient information to promote efficient markets, the participants would fear a loss of privacy. In response to this fear the database can contain indicators of performance rather than the raw data. To make this point explicit consider the health of a job candidate for a job of finite duration. A pertinent input in evaluating alternative candidates is the health forecast over the time span of the contract. To resolve the conflict between the need to keep personal data private and market efficiency, assume that entrepreneurs have developed several competing programs that read the pertinent data from the subject's medical record and forecast health. The potential employer is entitled to the forecast but not the raw data. To protect individual's medical records the holder of the medical record would run the evaluation programs as a service and would release only the output.

In order to compete for employment, most individuals would release the summary statistics to the job market database managers. The right of privacy beyond the common law protection directed at the media would be the right not to be compelled to release information for database analysis when no recognized study had demonstrated the pertinence of that data. Some information such as race, creed, or sex might continue to be restricted by law.

Ensuring the individual's privacy from the collection of extraneous data of possible subjective interest benefits the individual only if that right is enforceable. Simply passing laws that database managers cannot collect extraneous data would be difficult to enforce as Fourth Amendment protection limits searches to find extraneous data to probable cause. One scheme which would make the collection of extraneous information much more expensive and would help to ensure individuals released only required information in credential checks is the new cryptographic invention known as a blind signature[13].

Another safeguard which would help to make database operations self-enforcing is to grant subjects information rights to the decision process. To illustrate this point, consider the processing of a loan application by an expert system. Suppose the subject was granted the information right to know what analytic decision rule would be used to make the decision. Using his personal computer he could find out the amount the financial institution must loan him. Failure of the bank to grant the loan would be grounds for possible legal action on the part of the subject against the bank for using another criterion, or extraneous data.

Likewise, if the various analytical screening functions used to make decisions concerning jobs and other human concerns were in the public domain, the participants could police the process themselves without a great deal of government intervention. To ensure equal opportunity the state would require that all job openings be announced by being placed in the appropriate file in the social nervous system and that the screening functions be based on competency and not personal prejudice.

To provide decision makers freedom to act, the employer would be allowed to use for analytic screening any inputs and criteria which had been demonstrated significant for predicting performance. To inhibit lawsuits, he would prepare a short statement of references. The onus would then rest on the challenger to demonstrate that the evaluation criterion or inputs were not significant. If employment decisions occurred primarily through the database screening, then social standards for nondiscrimination would be quickly established.

Many decisions would have both an analytic aspect, in the sense of reducing a large pool to a small set of final candidates, and a subjective aspect, which would be the evaluation of the final candidates. The participants could police the analytic part using the rules for extraneous inputs and valid criterion. In such a competition the losers would know the evaluation criterion and could independently compute their rankings with the winning set. Thus, to satisfy due process, employment decisions would increasingly be based on merit. In time, this merit-based system for making employment decisions will become essential, as international corporations will have to guarantee due process for competitors of vastly different cultures.

Two other safeguards protect individuals and legal entities from time delay in the release of information. To enable the stockholder to analytically evaluate his principal agent and the voter his government official, the right to learn implies information rights for accountability, and as models to analyze principal agent behavior grow in size and complexity the demand for inputs for accountability will grow accordingly. This right conflicts with other informational concerns. For private firms, one way in which this conflict occurs is with trade secrets. To resolve this conflict, some accountability information would have to be released with a time delay. The faster the rate of technological change the shorter this time delay. Voters would have similar information rights to analyze the behavior of government officials. To facilitate decision making by public and private officials, the release of information on decisions not subject to an open meeting policy or disclosure policy would occur after the fact, since the principal protection from day-to-day interference in decision making is the time delay in the release of information.

A second safeguard for individuals is the right to deny physical and electronic access to their communities, a right which was discussed in the chapter on the community. Given a much more open political economy, more explicit defenses against possible information abuse must be constructed. As was pointed out previously, the town has the right to act as an electronic unit in purchasing services. This creates a mask between the members inside and the institutions outside. Secondly, households in town would have the right to limit and electronic access. For example, an individual, if he desired, could determine the number of a caller before deciding to respond. He could program his computer to filter incoming calls to respond to only those he wished to respond in person. Similarly, an individual could defend himself against unwanted ads by simply filtering them out. The right of limit access would provide the individual with the right to be left alone if he so desired.

Scientific information policy

The goal of information policy for science is to ensure that the society acquires empirical knowledge of all phenomena as rapidly as possible. As discovery is the basis for invention and innovation, the scientific information policy aims to promote a rapid rate of invention and innovation. For most scientific activities leading to invention there is no need for an explicit information policy, since there is no conflict between scientific curiosity and other social concerns. But for many scientific activities leading to innovation there is a fundamental conflict between scientific curiosity and the desire for privacy and trade secrets.

These conflicts between curiosity and privacy currently restrict the advance of empirical science concerning all types of human behavior. In some areas of social sciences such as conditioned learning experiments in psychology and agricultural economics controlled experiments are the rule. But in most areas of the social sciences and business disciplines the empirical research must discriminate between alternative hypotheses on the basis of observations of actual political economic behavior. In social systems the principle impediment to observing behavior is that the observed subjects vigorously oppose being observed. What this means is that empiricists in testing an hypothesis must make ingenious use of the available observations rather than collect a data sample best suited to test the hypothesis. Numerous social and political economic phenomenon are simply not observable in any systematic fashion.

The difficulty of obtaining observations has created a theoretical bias in many disciplines of social science and business administration. Currently these sciences, such as economics, have developed mathematically rigorous theories of behavior, but at the same time can not accurately explain the behavior they are ostensibly studying on a moment by moment basis, that is real time. For example, there are no simulation models which will simulate the behavior of a firm in real time to an accuracy of even one significant figure.

The academic incentives of publishing in prestigious journals to some extent impede rather than promote scientific advance, inasmuch as the criterion for publishing in the respected academic journals is, unfortunately, mathematical rigor and sophistication rather than the ability to explain the phenomena under study in real time. To be sure, the bias towards the abstract mathematical manipulation of models is useful as a devise for ranking the mathematical ability of academics.

But in fact these academic incentives tend to create mathocracies rather than sciences explaining natural phenomenon. This is not to degrade the efforts of the many conscientious and able scientists who labor to extend the empirical contents of their respected disciplines. Rather, the problem is that a major conflict exists between generating observations which are efficient at discriminating between alternative theories and concepts of privacy and property rights.

To ensure a rapid rate of discovery for innovation, the right to learn needs to be defined to provide observation rights to all phenomenon. The minimum scientific observation right is a representative sample of observations on any phenomenon. Because the study of behavioral relationships does not require a knowledge of the identity of the subjects in the sample, the compromise with privacy is to delete labels in scientific datafiles. Observations involving legitimate trade secrets like military secrets would be released with a time delay.

To alleviate fears of abuse of data and to promote accuracy in observations, data collected for scientific purposes must not be used for administrative purposes. This means that scientific records should not be subpoenaed to reveal individual information, although the methodology of obtaining and using such records could be challenged. In this regard the incentives to promote accuracy are more important than the occasional discovery of illegal activity. Scientific records would, however, be used for policy purposes such as performing studies which might indicate the need for future changes in laws or procedures. To preserve the separation of scientific data from administrative data, agencies with no administrative functions should collect scientific data.

Private agencies collecting scientific data would likewise be regulated to ensure due care in preserving privacy. The use of scientific data for private operational decisions would also be prohibited. Scientific data could only be used privately to construct better decision functions, and if the new decision functions were superior to the old, they might create an operational demand for new inputs from all participants in a decision process. Until this happened, however, even if decision makers knew the identity of subjects in a scientific sample, they could not use the data for decision purposes. For screening functions in the public domain this separation of scientific and administrative data would be enforceable.

Given the proposed safeguards, scientific information policy can be used to greatly increase the empirical content of sciences and applied disciplines concerning political economic phenomenon. First consider the collection, use and release of data collected by government. Currently government at all levels collects information on economic agents for administrative purposes. Since the computerization of government records in the 60s, administrative agencies and researchers outside government have made increasing use of administrative data for statistical studies. This data has defects for studying behavior as the purpose of the data collection is for compliance with administrative law without consideration of issues of statistical methodology. Also inconsistencies in data collection and the Privacy Act and other legislation inhibit the creation of behavioral data samples from many administrative datafiles. Finally, researchers outside government can rarely obtain disaggregated business data which the government has collected.

Because of proposed safeguards, the collection and dissemination of administrative data could better serve scientific purposes. With the proposed distinction between administrative data and scientific data, the combining of data from many sources for statistical purposes should be encouraged. The quality of such composite data can be improved by greater attention to statistical considerations in the collection of administrative data[14]. Also as informational society advances, government agencies, to fulfill their legislative mandates, will increasingly collect data based on good statistical methodology to understand the behavior under their own particular mandate.

Moreover, information collected by the government can be released to researchers outside government either immediately or with a time delay. In a constantly innovating society, the purpose of protecting trade secrets is to create incentives for constant innovation in order to keep ahead of one's rivals. The value to a firm of data collected by the government so that the firm may examine the behavior of its rivals is rapidly discounted. Over time the files of data collected about business behavior would be released in a disaggregated form.

This means that for administrative data series such as the SIC code aggregated data would over time be available at the level of the individual firm and product, and eventually all the internal information of all public and private institutions would become public. Administrative data concerning individuals would be released in the form of unlabeled representative samples. The major change this represents from current policy is the increase in collection of representative samples of behavior.

While data collected by government agencies could better serve scientific purposes, more steps need to be taken to promote discovery of political economic behavior. Currently much empirical research asks what hypotheses can be tested by the available data, not what the most important hypotheses are that should be tested. Scientific information policy should provide empirical scientists representative data samples in order to test hypotheses of any behavior and the compliance in supplying these data would be compulsory. The concept that the right to learn guarantees the right to obtain a representative sample will appear very threatening to many. And, indeed, given the current political design, the fear of abuse of power is well grounded in numerous empirical examples. With a more open information policy firms would fear their trade secrets would be stolen, and that adverse publicity might inspire the public to demand new government regulation.

The scientific observation right is intended to cover all phenomena; however certain practical considerations place some limits on how this right is to be exercised. First, the observer pays for the cost of observing. Second, this right is a right of the scientifically competent, since funding for observations will be provided by such agencies as the National Science Foundation. In competing for funding, the proposals for observations will be subject to peer group review. Other government agencies will have an interest in resolving hypotheses important for innovations or policy, while private foundations would fund more controversial studies.

While professionals and professional societies could use the legal system to obtain the desired observations, in most cases they would not have to do so. In an industry-wide study, say of incentives versus performance of managers a professional society would, in order to obtain voluntary compliance, work with the trade association to soothe any fears of loss of trade secrets. Professional societies observing community behavior would have a similar interest in smoothing the situation with local officials, and much of the burden of data collection would be minimized by machine talking to machine.

Impact of information policy

Whether voters would favor the proposed information policy depends on its impact. Operational information policy would promote better decisions in general and would tend to vastly improve the decision-making capabilities of individuals. Scientific information policy would promote a faster rate of innovation. Voters would gain more from better decisions and a higher rate of innovation than they would lose in loss of privacy.

In current society the cost and power of information technology means larger institutions such as major corporations and government have much greater information access and processing capability than individuals. In current society individuals make most decisions based on intuition with little access to databases. Operational information policy would greatly reduce the asymmetry between individuals and institutions. With the expanding social nervous system database entrepreneurs would create numerous databases for individual decision making. These database entrepreneurs would have the resources to obtain negative information such that their clients would make more informed decisions. With much more powerful home computers and distributive processing through the social nervous system, individuals would be able to analyze their alternatives systematically.

Scientific informational policy would stimulate innovation by greatly improving the empirical content of social sciences and the allied business disciplines. Researchers would have information archives for the study of empirical issues of society and the political economy. These archives would consist of observations from experiments, sampled observations of social and political economic behavior, data files constructed from administrative data, and finally public and private data and documents with a time release schedule. These archives would be accessible to researchers. With this access to better measurements, all disciplines studying social and political economic issues would be able to more rapidly discriminate between hypotheses. Also, much more effort would be made to collect data to discriminate between important hypotheses than currently.

Furthermore, the impact of scientific information policy on disciplines studying social and political economic behavior would be more than just better hypothesis testing. Consider, for example, the field of economics. Current economic data makes the study of economics in real time very difficult. As microbinics advances, however, the time interval between economic measurements will decrease. At the same time, the scientific right of observation will enable economists to study economic behavior with approximately continuous measurements. As a consequence, these measurements will accelerate the trend to study economic behavior from the perspective of dynamic models in real time.

With constant change in technology all political economic agents must constantly adapt to change. With information policy that supports the methodology of empirical science, all agents, both private and public, could harness the methodology of science to innovate better decision rules, incentive systems, and organizations to cope with an environment of constant change.

To illustrate this point consider the decision rules for preliminary screening in databases or for such activities as granting credit. Because of the requirement that decision rules be placed in the public domain, the scientific information policy would create enormous incentives for systematic study of decision rules. As better decision rules are discovered they would be quickly implemented and the requisite data placed in operational databases. In this process researchers would use much more extensive scientific databases based on a selected sample of unlabeled subjects to test theories. Database managers would then quickly add to the labeled operational databases those variables found to be significant for decision making.

The greater emphasis on empirical science would promote innovation in general in two ways. First better hypothesis testing would tend to eliminate spurious theories from consideration as the basis for innovation much more rapidly than currently. Second with better hypothesis testing the predictive capacity of theory should increase. Thus innovators would be in a much better position to analyze the projected impact of possible innovations without having to subject them to an empirical test.

The public acceptance of a more open informational policy depends on a majority of citizens gaining from the change. Given a system of social inheritance, the motivation of all citizens is modified. As he inherits his share of the capital stock, everyone simultaneously plays the role of consumer, producer and public. As a consumer, the individual would like to acquire goods at the competitive market prices. As a producer, however, he would want the highest rate of return, and as a citizen he would like an effective government. Not only are these goals in conflict, but their resolution will change as new knowledge is acquired about the environment, more efficient production technologies, and consumer needs.

What individuals gain from operational information policy is better decisions in each of their roles and better information to resolve the conflicts between their roles. Scientific information policy promotes a higher rate of innovation. What individuals would lose in selling their labor services in the marketplace, however, is control over the release of pertinent data, and they would, also, be subject to more census type compulsory data samples for scientific studies. Firms would lose a competitive advantage which is now based on poor flows of information. On balance between the positive and negative aspects, most people would be better off as such a political economy would be much more competitive in international competition than the current political economy with a more restrictive information policy. As we shall see in the next chapter, the proposed information policy leads to better government.

Notes and References 7

1. The plant closing law was originally part of a general trade bill which President Reagan vetoed in the spring of 1988. The final version, which became law without President Reagan's signature, is the Worker Adjustment and Retraining Notification Act, PL 100-379

2. Taylor, T., 1955, *Grand Inquest*, (Simon and Schuster: New York)

3. Wilcox, Clair, 1960, *Public Policy Toward Business*, (Richard D. Irwin, Inc.: Homewood, Illinois)

4. Gellhorn, W., C. Byse and P. Strauss, 1979, *Administrative Law*, (The Foundation Press, Inc.: Mineola)

5. The information policy concerning various consumer credit legislation is contained in: Redden, K. R. and J. McClellan, 1982, *Federal Regulation of Consumer-Credit Relations*, (The Michie Company: Charlottesville)

6. OSHA mandated the disclosure in 1983 with the publication of the Hazard Communication Standard.

7. See reference 4.

8. Stevenson, R. B. Jr, 1980, *Corporations and Information*, (The John Hopkins University Press: Baltimore)

9. Prosser, W. L., 1960, Privacy, 48, *California Law Review*, 383. Professor Prosser's ideas are incorporated into Restatement (Second) of Torts.

10. For a summary see: Freedman, W., 1987, *The Right of Privacy in the Computer Age*, (Quorum Books:New York) 11. For a discussion of the conflict between efficiency and privacy in the sense of being able to withhold information see: Posner, R., 1981, *The Economics of Justice*, (Harvard University Press: Cambridge)

12. Chrysler attempted to block the release of employment data under FOIA. See: Chrysler Corporation V. Brown, 99 Sup. Ct. 1705 (1979). The economic incentive for this action in discussed in reference 4.

13 An individual does not reveal his or her identity electronically when using a blind signature. See: Chaum, David, 1992, Achieving Electronic Privacy, *Scientific American*, Aug, pp 96-101. This raises the cost of obtaining the information by other means and in many cases the cost would be prohibitive.

14. Staff, 1980, Statistical Policy Working Paper 6, Office of Federal Statistical Policy and Standards, US Department of Commerce

Chapter 8

The Federal Government

Introduction

As was pointed out in Chapter 2, the overall performance of the political economy has become critically dependent on the performance of a large government. To name just a few examples: economic performance depends on the performance of government monetary and fiscal policy, business invention and innovation performance depends on government promotion of research and development and education, and the status of the environment depends on the implementation of environment policy. Yet as was also pointed out in Chapter 2, government has developed serious problems in its ability to make a good estimate of the common weal and for the most part has not developed an strategy for implementing innovations appropriate for its expanded role.

The purpose of Chapter 8 and 9 is to propose fundamental changes in government in order to achieve a high performance government for the projected political economy of the mid 21st century. In this chapter changes in the system of checks and balances will be proposed in order to achieve a much better estimate of the common weal[1]. In the next chapter the implementation of much greater decentralization will be proposed in order to achieve a higher rate of learning.

A starting point to propose changes in government which will result in a better estimate of the common weal is to review the defects discussed in Chapter 2. As the scope of government has expanded, increasingly specialized committees and subcommittees propose legislation. Moreover, because many government actions have an uneven distribution of costs and benefits, these committees have an intensity bias, that is, they tend to promote desires of groups who receive disproportionately large costs or benefits. The specialized legislature tends to estimate the common weal as a collection of goods for various constituencies. Considered individually many government actions lack general benefits, farm subsidies, for example, lack benefits for consumers. Also, the estimation of the common weal does not achieve even approximate consistency. One example here is the exclusion of the oil industry from the superfund cleanup. Finally, because there is no direct relationship between the costs and benefits of most government programs, voters do not have strong incentives to demand efficiency in government. In short, government actions frequently lack consistency, general benefits, and efficiency.

One possibility of reducing these shortcomings is to drastically curtail government to the idealized definitions of limited government of the 18th and 19th centuries. Such an effort would transfer much of the effort of estimating the common weal from the collective action of government to private action of individuals. The author assumes such an effort is not possible for the foreseeable future. While some concerns might decrease, for example the military if there is a relaxation of international tensions, the overall size of government is unlikely to decrease significantly as other concerns are likely to increase. For example, the concern for the local, national and international environment is bound to increase given current events such as the demise of forests due to acid rain, the collapse of whole ecosystems which has had the greatest loss of lifeforms since the demise of dinosaurs and the depletion of the ozone layer which has created an ultraviolet radiation hazard. Therefore, the author assumes the task of proposing changes to improve the estimate of the common weal of a government which has a large role in the political economy.

Given the vastly improved technology for communication, some idealists might wish to improve large government's estimate of the common weal by replacing representative government with a direct democracy[2]. As the capacity of the communication system expands, the feasibility of implementing a direct democracy increases. The real issue, however, is whether the implementation of a direct democracy would in fact improve government performance. Voters, after all, are bounded rational beings with limited resources to invest in deciding between political alternatives. And as government grows in complexity, the amount of resources voters must spend on making informed choices also grows. Even if databases were prepared with complete analysis to aid the voters' efforts, an increasing amount of time would be required for voters to understand the analysis. In as much as legislators currently rarely understand the details of legislation outside their specialty, it is unlikely that the average voter would trouble himself to be informed on issues other than those which directly affected him. Consequently, direct democracy is not likely to ameliorate the problems of intensity bias.

Hence, the subject of direct democracy will not be pursued in this book for two reasons. First, as has been pointed out, direct democracy is unlikely to improve the estimate of the common weal. Second, it is assumed that the more complex government becomes, the more the voters, in order to reduce their expenditures of resources in political decision-making, would prefer to delegate this task to specialists, namely their representatives.

Therefore, the agenda is to propose changes in the system of checks and balances of our representative government in order to promote an estimate of the common weal which has the properties of general benefits, consistency and efficiency. Originally, the founding fathers incorporated a system of checks and balances in the Constitution to promote an estimate of the common weal with the properties of maintaining a limited government and property rights. In this chapter the author will propose numerous institutional changes in order to create a new system of checks and balances which will result in an estimate of the common weal which has the new desired properties. To motivate the proposed changes we need to reconsider why the current governmental design only weakly promotes the properties of general benefits, consistency and efficiency. These properties are best determined by teams of experts. For example, the general benefits and efficiency of a particular environmental policy is best determined by a team of economists, biologists, engineers, and other experts. It should be noted that the problem is not that teams of experts refuse to perform such analyzes. Over the past several decades there has been an increasing number of articles in professional journals analyzing all aspects of governmental policy.

The problem is that even when most experts agree that a particular governmental policy is seriously flawed, there is no current mechanism other than majority vote in the legislature to change a law. And the vested interest benefiting from a bad law can easily block change in the current specialized legislative committee system. Moreover, given the complexity of governmental issues voters are unlikely to demand action from their representatives unless the defect has precipitated a crisis or directly affects them Hence, serious flaws in governmental policy can take decades to correct.

In order to create a more rapid mechanism to correct serious flaws in governmental policies a professional check, which hereafter will be called a *professional review* is proposed. This professional review would be a formal governmental activity integrated into the system of checks and balances in order to create powerful incentives for elected and appointed officials to incorporate general benefits, consistency and efficiency in all their acts without the need of constant professional reviews.

As modifying the system of checks and balances will require Constitutional amendments, the proposed design must consider keeping the proposed changes to a minimum to ease the problem of obtaining voter approval. To the extent that the proposed changes might find public favor, the assumption is that they would be implemented by constitutional changes over the course of the next half century. By this time the forecasted technological changes described earlier should be in existence. The discussion, moreover, will focus only on changes to the present government rather than providing a complete exposition. The order in which the three branches are discussed corresponds to the magnitude of the changes proposed for each. The order will be the judiciary, the legislature and he executive.

Judiciary

The first step in integrating the professional review into the system of checks and balances is to explain the concept. A professional review is a judgment by a group of experts from the relevant disciplines whether a legislative or administrative act has the desired properties of general benefits, consistency and efficiency. To keep such professional reviews from exceeding current knowledge in measuring these properties, the experts making such judgments would not be expected to render their personal opinions, but rather they would use their expertise to determine whether the analyses published by professionals in referred journals were in agreement or disagreement on the issue under consideration. To declare an act unconstitutional the experts would have to find general consensus that the act lacked at least one or the desired properties.

The consensus criterion is imposed to limit the ability of any individual, for example a very liberal or a very conservative economist, to impose his personal opinions on government. Such a limitation is considered necessary in order to gain popular support for the concept. But because experts in the various disciplines are likely to have widely varying opinions on many government acts, it might seem that the consensus criteria in the professional review makes the professional review so weak that it will have little effect. Nevertheless, it will be demonstrated that properly integrated into the system of checks and balances, the professional review would promote a much better estimate of the common weal.

There are many ways in which a professional review might be integrated into government. To avoid adding to the complexity of an already complex government, the professional review should be integrated into one of the existing branches of government. In order to insulate the professional review from the short-time-horizon concerns of reelection politics, the professional review should not be placed in either the legislative or executive branch of government. This leaves the judiciary.

Let us now consider two ways in which the professional review might be implemented in the judiciary. First, one or more special courts could be set up as science courts to make professional reviews. Second, experts in various disciplines could be added to the bench of the current system of courts to enable the current courts to make professional reviews. We shall opt for the second approach by expanding the staffing of the judiciary to include judges-in-fact as well as judges-in-law[3]. The former would be experts in various disciplines and the latter would be traditional legal judges.

An important reason for expanding the judiciary to include experts from various disciplines is a fundamental response to the rapid rate of discovery in all disciplines. Judges-in-fact are needed to ensure that people making judgments in informational society are competent to understand what they are judging. As knowledge advances and is applied to all human endeavors, the ability of a jury of one's peers or for that matter judges-in-law to make competent judgments of fact on the basis of intuition will decrease. Furthermore, the jury and judges-in-law will become increasingly less competent to evaluate disputes between expert witnesses.

Thus, expanding the judiciary to include experts of various fields is necessary to ensure that the judges understand the evidence in the case. Since a single person is unlikely to understand all the issues of a complex case, a committee of experts spanning the law and facts to be presented in the case would increasingly judge cases. The assumption is made that jury trials would gradually be displaced by trial by expert committee. The proposed precedent for an expert committee trial is that if either party intends to use expert witnesses, then the case would be judged by a committee of judges competent to judge both the law and the expert witnesses[4].

Assuming the current heavy case load of the judiciary is unlikely to decline in the future, expanding the bench to include the new professionals would improve court performance. The Supreme Court would be expanded to, say, 15, together with corresponding enlargements to the courts of appeals and the district courts. The number of professionals from each profession should correspond to the number of cases requiring expertise in the respective profession. For example, physicists would be required to judge nuclear issues, chemists and biologists to judge environmental issues, medical doctors to judge medical issues, and psychologists to judge sanity issues. As a great many legal issues involve property, a large contingent of judges-in-fact would be professionals from business administration fields and economics. The number would undoubtedly vary over time and should be established by slowly changing customs rather than by Constitutional amendment or statute. The distribution in the number of judges from each selected profession would probably be biased towards the prestigiousness of the profession; nevertheless, over time the composition of professionals on the bench would roughly reflect the knowledge necessary to judge disputes.

In revising the judiciary, an important concern is the appropriate length of appointment for members of the bench. To promote an independent judiciary the framers of the Constitution made appointment to the federal bench for life. But as lifetime expectancy has increased by several decades since 1790, does lifetime appointments still promote judicial performance? Maintaining lifetime appointments would promote independence and would also help to maintain the judiciary centered in the political spectrum, as each President would appoint only a fraction of the judiciary. This is desirable in order to achieve orderly change in law as knowledge advances.

On the other hand, an argument for shorter appointments is that they would ensure that judges' knowledge not become obsolete during office. So to achieve better performance with minimum changes to the current system, a compromise between these two options would be a requirement that federal judges retire at the age other professionals in their fields retire. If this criterion were applied in current society, judges, like university professors, would retire at seventy. This would promote more orderly change, especially in the Supreme Court, as judges would not have the option of trying to outlive a President who maintains a different political philosophy.

The next issue to consider is how technological advances could be employed to improve judicial performance. Currently the district and appeals courts have jurisdiction over specific physical districts. Within these physical districts the ability of judges to specialize in particular types of cases is limited. As knowledge advances, greater specialization would result in better judicial performance. The federal bench could become much more specialized if judges presided over cases throughout the entire country rather than the current physical districts. With advances in communication such a court system could be effectively administered through teleconferencing. The system of courts would thus become a national system of teleconference centers, and the participants would simply use the nearest center. Judges would specialize in areas of law and would hear cases in their specialty nationwide. With teleconferencing and national case loads the judges would develop, through on-the-job training, a better understanding of the law and facts of their specialty than is currently possible, because of the much narrower range of cases over which they would preside.

A more specialized judiciary, however, creates a potential problem in maintaining the independence of the judiciary from the other two branches of government. In a more specialized system of courts, economic interests strongly influenced by a particular judge would have a great interest in determining the next appointment. If judicial appointments were for narrow specialties, the judiciary would accordingly tend to suffer the same type of problems that independent regulatory agencies currently suffer. The most concentrated economic interest group would have great incentives to appoint a friend of the interest. To prevent this possibility, the power of appointment to particular cases should be solely the function of the judiciary. To assist in this process, an assignment committee would assign cases to judges based on the knowledge requirements in law and fact for each particular case. The bench would elect this committee with the appointments of each President electing one representative. The committee would elect a chairman each year from its elected members. For a large number of cases of a particular type, such as bankruptcy, the assignment committee might appoint a specialized court to sit in permanent session. These specialized courts would hear cases nationwide not just in a physical district. Moreover, to maintain independence from the other branches of government the assignment committee would reassign judges sufficiently such that the various interest groups in society would only have a general rather than a specific interest in the appointment of a judge. In addition, the review process would tend to limit the ability of outside interests to control specialized courts.

Assuming the judiciary would be able to maintain its independence from the legislature and executive, the quasi-judicial activities of the administration should be transferred back to the judiciary. Such a move would promote the original separation of powers between the judiciary and the other two branches. As was pointed out in Chapter 2, with the growth of the bureaucracy especially the independent regulatory agencies, the administration has acquired quasi-judicial activities, that is, adjudicatory judgments by administrative officials. One example is an administrative law judge of the National Labor Relations Board deciding an unfair labor practice case. Under the current system, administrative agencies tend to represent the interests of the group with the most concentrated interest in the agency and policies tend to shift abruptly with a change in the President. Long term regulatory performance should improve if control were passed to independent slowly changing judiciary competent to judge both matters in law and matters in fact .

The final aspect of reorganization of the judiciary would be to expand the scope of the judiciary to cover all new forms of expert dispute resolution. Given the costliness of traditional trials, newer forms of settling disputes such as small claims court and binding arbitration have been devised[5]. Advocacy trials may not be the best mechanism for settling disputes involving scientific forecasts of future events. Undoubtedly, over time other forms of dispute resolution will be proposed and tried. All forms of dispute resolution by experts would become part of the new expended judiciary.

Having considered the reorganization of the courts, let us now consider how the courts would process professional review cases. As is the precedent for judicial review cases, the judiciary would, to avoid interfering with the legislature and executive, only consider professional review cases brought before the court by an interested party after the legislature or the executive had acted. In that it is now generally accepted that the Supreme Court interprets the Constitution, the Supreme Court would have the last word in a professional review. For the professional review to create incentives for the legislature and the executive, the court must operationally define the meaning of the abstractions: consistency, general benefits, and efficiency. Given finite knowledge, of course, bounded rational judges will never be able to create a single set of definitions appropriate for all conditions and all times. Therefore, in judging cases the court would have to create a set of precedents or rules of thumb which would apply to categories of cases. These precedents would provide the legislature and the executive with useful information, as they would tend to define those actions which clearly would not pass a professional review.

To improve the adjustment of precedents to changing knowledge and conditions the analysis supporting a precedent should include a complete analysis of the impact of the precedent on society[6] in addition to the traditional legal analysis. The legal analysis would be performed by the judges-in-law and the social impact analysis would be performed by judges in fact . With greater specialization, judicial opinions will improve thereby providing better adaptation of precedents to changing conditions and as knowledge advances and measurements improve, the requirements of a professional review would become more exacting.

Now let us consider the desired properties of general benefits, consistency, and efficiency in greater detail. A minimal requirement for general benefits would be that all legislative and administrative acts be based on a complete analysis demonstrating that the proposed action is in the common weal. If the standards of this analysis were set as a demonstration of conclusive evidence, very few government actions would pass the general benefits test. Because estimates of the common weal differ widely between liberals and conservatives and because there are competing theories for most social phenomena, critics could successfully challenge almost every proposed government action. In order to provide a much greater range of government action, a lesser standard is proposed[7]. For legislation or administration to be constitutional, the analysis would only have to be considered correct by one of the competing theories. Thus a President, in promoting his legislative program, would only have to ensure that the legislation was correctly analyzed by theories most congenial to his political beliefs.

While this criterion might seem very mild, it would have a great impact over time. As knowledge advances some theories are rejected as invalid by most practitioners in a profession. Once this happens any legislation or administrative action based on the rejected theories would fail to pass the criterion of general benefits. In the thirties much regulation of industry was enacted to obtain legal cartels in various industries. By the sixties most economists had rejected the theoretical basis of this type of legislation. Given a professional review, the deregulation of airlines, banks, and trucking might have occurred through the courts about ten years prior to the deregulation legislation of 1980.

Over time precedents would be established for the amount of evidence needed to establish general benefits in proposed legislation and administration. Very few laws have equal impact on all citizens; many, for instance, provide immediate benefits to a small group and indirect benefits, or even harm, to society as a whole. One example of this is legislation regarding basic research from which scientists enjoy the direct benefits and society enjoys the indirect. The principle which should be employed in making a general benefits test is that the smaller the subgroup which receives the direct benefits, the greater should be the evidence of general, indirect benefits. For example, at the very least most theories would have to concur that general indirect benefits existed. The precise evidence required to prove indirect benefits would change, of course, as knowledge and the ability to measure benefits advanced. For instance, by this measure expenditures on much basic research would pass the criterion, but expenditures to preserve homes on barrier islands along the Atlantic and Gulf coasts would not.

An exception to the general benefits rule is the case in which the cost of a service must be paid for by the recipients of the service. Currently licenses and special taxes on fishing and hunting gear pay for much of the government efforts in research and development of better fishing and hunting. This criterion would necessitate that the full cost of hunting and fishing promotion be paid by the recipients. Government services such as search and rescue would also be covered by this criterion. As microbinics advances the ability to charge recipients for such special services such as this will increase. Many programs have differential general benefits between the nation and the state where the facility is located. For example the super collider will advance particle physics worldwide, but will only provide local jobs for construction and maintenance. In such cases, the financing for such programs should be split between the higher and lower levels of government.

A special case of general benefits is the criterion of simplicity. The idea behind the goal of simplicity is that unnecessary complexity in government laws or programs is the fertile ground for creating unwarranted privileges. The classic example here is the old tax code, in which incredible complexity created numerous opportunities for special interests. In legislative and administrative acts, therefore, the burden of proof of performance should be placed on those who wish to replace a simple system with a more complicated system. For a more complicated system to replace a simpler system, there should be general agreement among the respective experts that the more complicated system is better. This challenge would serve to keep legislation and administration as simple as possible.

The second property of consistency provides a professional criterion for analyzing the various subgoals of legislation and administration. Given the current, compartmentalized nature of legislative committees and administrative departments, serious inconsistencies in government actions arise. Moreover, voter comprehension of government is insufficient to demand consistency as a performance criterion. A classic example of this inconsistency is promoting tobacco through subsidies while at the same time mandating a hazard warning on cigarettes packages. The treatment of tobacco, alcohol, and drugs, for instance, is inconsistent with respect to the social harm of these substances. Another example is the inconsistencies in risk management by the government[8]. Government regulation places much higher standards on preventing new risks than in reducing old risks. Old risks are covered by standard setting in which the industry's operations do not have to be modified until the regulatory agency issues a standard. In contrast, new risks are screened meaning the industry must obtain advance permission before beginning operations. For example, the Environmental Protection Agency sets standards for producing old chemicals, but screens the production of new chemicals. As a result, two inconsistencies arise. Generally new activities must pay for the cost of screening while the regulatory agency pays for the cost of setting standards. In addition, the standards for new activities are frequently much higher than those of old activities. These inconsistencies have negative social implications in that new technological activity is inhibited and insufficient effort is directed at reducing the risks of old activities. The precedent for inconsistency should be that if an inconsistency can be demonstrated to have negative social consequences, the court would provide the legislature a legislative session to resolve the matter before imposing a court mandated solution. With a challenge of consistency the tobacco subsidies would probably be eliminated. The challenge of consistency works to promote action towards more overall consistency in legislation and administration.

The third property of efficiency would provide a criterion for numerous challenges to legislative and administrative acts. An example would be legislative interference in attempts by the administration to close inefficient facilities such as military bases. The efficiency property would give the administration a mechanism to pursue policies which were efficient nationwide, but which might cause a reduction in employment in a particular congressional district. In cases where the technology of some government operation has fallen behind the corresponding operation in the private sector, vendors of the new technology could challenge the inefficient government practice. Their incentive, of course, would be to promote sales of their new technology. This challenge would also promote the use of more sophisticated techniques to provide the correct level of public goods[9].

The requirement that the legislative and administrative acts have the desired properties of general benefits, consistency and efficiency could engender an extremely large number of professional review cases. To keep from being overwhelmed by such cases, the judiciary would adopt a precedent that professional review cases must be fought out in professional journals prior to submission to the courts. Thus, a brief to initiate a professional review case would have to be based on referred research published in a journal with a track record in previous professional review cases. Currently there are several reputable journals which specialize in policy analysis. The precedent for published research as a basis for initiating a professional review case would spawn numerous new journals. Undoubtedly, the bastion of market economics, the University of Chicago would immediately create one or more professional review journals explicitly for the purpose of pruning governmental excesses. Other universities would follow with additional creations.

The precedent that a professional review be based on published research would greatly increase the amount of policy research performed. But because of their heavy case load, the judiciary would be able to hear only a small component of potential professional review cases. For the professional review to have a major impact on the estimate of the public weal, the professional review must be integrated into the system of checks and balances to create powerful incentives for elected and appointed officials to incorporate the properties of general benefits, consistency, and efficiency into proposed legislative and administrative acts prior to enactment.

Legislature

In this section numerous modifications to Congress will be proposed which would improve the performance of the legislature in informational society. The estimate of the common weal would be improved if legislators had incentives in creating legislation to carefully comply with established precedents for general benefits, consistency and efficiency. Also as will be discussed, the performance of the legislature in creating legislation, conducting administrative oversight and solving constituent problems would improve if the task of creating legislation were separated from the other tasks of the legislature.

Let us first consider the problem of creating incentives for legislators to carefully comply with established precedents for general benefits, consistency and efficiency in creating legislation. We assume most legislators enjoy the power and prestige of office and seek long term political careers. Consequently, legislators in pursuit of power have strong incentives to take whatever action is required to ensure re-election. To make the professional review a re-election issue, a legislator in sponsoring a bill or amendment would assume the responsibility that the bill would pass a professional review. And the current intensity bias in the campaign financing of elections would be greatly reduced by publicly funded elections with fixed limits for each level of office. Currently, the unlimited campaign financing by political action committees, PACs, and private contributions gives a disproportionate influence to concentrated interests in elections.

With fixed amount, publicly funded elections, the legislator wishing to remain in office would have to be prepared to face an opponent with equal funding. Moreover, the voters with limited resources to carefully analyze issues would focus on simple measures of performance which they understood. A negative professional review becomes a simple measure of failure to perform which an enterprising opponent would quickly bring to the attention of voters. A negative professional review on the grounds of lack of general benefits is a clear sign that the incumbent has been captured by special interests. A negative professional review on the grounds of lack of consistency or efficiency is an indicator that the legislator may not be competent. Consequently, legislators wishing a long political career would subject sponsored bills to much more careful scrutiny than is the current practice.

To make the incentives for analysis even stronger, we must reconsider the concept of a legislative district[10]. Currently legislative districts are physical districts. This means that legislators who have a major impact on national legislation, such as the chairmen of powerful committees, are not responsible to the national electorate. Moreover, the local voters who decide the re-election fate of a powerful national legislative figure may have quite different interests from those of the national electorate.

To encourage accountability to all voters, legislators would be elected in national elections. Given the broad scope of government and the complexity of political issues legislators should run for legislative specialties which are defined as particular areas of legislation–for example, finance, social programs, or defense. Prior to each election Congress would specify the legislative specialties which would be fixed until the next election. Legislators running in one of these legislative specialties would be guaranteed committee assignments in the area specified.

The proposed national elections would shift the focus of re-election politics; each legislator would have to consider a majority in terms of a national constituency rather than a small physical district. While a charge of special interest would be an asset in a district where the special interest was the majority, such a charge would be far more damaging in the perception of a national constituency. Consequently, nationally elected legislators would necessarily be much more concerned about general benefits than would legislators elected from geographical districts.

To implement a Congress elected by legislative specialties, the limited resources and bounded rationality of the voter must be considered. Given the limited resources that the voter can devote to election decisions, the number of legislators which the voter should have to evaluate in an election should be quite small. We will assume that the voter should not have to elect more than seven at any one time. Under a system of annual elections and six-year terms of office, Congress would be composed of only forty two members. But even with the decentralization suggested in the next chapter, we assert that such a Congress would have too few members to conduct the legislative tasks of informational society.

Thus, in order to obtain a larger Congress, each voter would elect six members each year for national legislative specialties and one additional member from a geographic district[11]. We will stipulate that two hundred legislators elected on the basis of geographic districts should be adequate. Each state based on population would be allotted its share of geographic district legislators whose districts and length of service would be specified by the respective state. Congress would thus consist of 237 members, one vice president, 36 nationally elected senators, and 200 congressmen elected from the various states.

The final change to improve the estimate of the common weal is that only senators would have the power to sponsor legislation or amendments. This power is because senators with national constituencies would be more concerned about general benefits than congressmen with localized geographic constituencies. But the hazards of re-election politics from negative professional reviews could create a chilling effect of sponsoring any type of legislation at all. For example, currently senators and congressmen can have long successful careers by specializing in promoting constituent relations with their government and rarely if ever sponsoring legislation of any type. Thus, under the career hazards of a professional review, few, if any, senators might be willing to sponsor legislation.

To create stronger incentives for senators to sponsor legislation, the functions of the two houses would be specialized. The function of the Senate would be to sponsor all legislation and the function of the House would be to conduct administrative oversight and to address constituent relations with government. Given this split, the Senate would focus on sponsoring new legislation and modifying the President's legislative program, a feature of government originating with F. R. Roosevelt and likely to continue for the foreseeable future. Before sponsoring legislation the senators would subject the legislation to intense analysis in order to ensure themselves it would pass a subsequent professional review. As knowledge accumulated, the Senate would periodically analyze existing legislation for the three desirable properties..

Nevertheless, the professional review would not suddenly shift the re-election incentive of each senator from that of maximizing the benefits to his constituent groups to the lofty national purpose of promoting general benefits, consistency and efficiency. Rather, the senator would be interested in performing analysis to ensure that his promotion of his constituent's interests lies within the precedents established by previous professional reviews. The senators would thus have a strong interest in establishing their own professional review mechanism to greatly reduce the risk of professional review challenges in the courts. Senators themselves are likely to be lawyers, businessmen, scientists, and ex-celebrities such as basketball players, astronauts, and actors. While some of the senators would be specialists capable of performing their own analysis, most senators would delegate the analysis to their staffs. The role of the senator, after all, would be to reach compromises within the constraints of the professional review.

In short, the professional review would change the process of creating legislation in various ways. First, senators would insist that the President, in submitting proposed legislation to the Senate, would also have to submit the careful analysis upon which the proposed legislation was based. In the Senate legislation would be submitted to a further two part review. First, public hearings with media exposure would emphasize the human aspects of the legislation and senators as media celebrities would make their traditional political statements concerning the purposes of the legislation. Second, a technical analysis of the legislation would be conducted by specialists presenting opposing analysis of the proposed legislation. These proceedings would generally be too technical to create a great deal of public interest. Staff members of the senators would do most of the work, with senators putting in an occasional appearance, especially if the hearings became media events. These technical hearings would generally be teleconferences between experts located throughout the country. To support technical analysis of legislation senators would have to greatly increase their technical staff.

The role of informational society senators who would sponsor legislation would be complicated. They would have to provide the electorate with the type of legislation which would appeal to a majority in order to prepare themselves for the next election. In sponsoring this legislation they would have to make sure that a majority of the Senate and House would vote for the legislation, and to obtain this majority they would have to make deals with with congressmen concerned with their local interests. In making the necessary compromises in this process, they would have to stay within the bounds of possible professional reviews. To compensate for such difficulties, the role of senators would be even more powerful than it is currently. In informational society senators would be celebrities who would enjoy considerable power with considerable risk.

In the new design, two aspects of the relationship between the Presidency and the Senate would change. First, the Vice President would become the leader of the President's legislative program in the Senate. He or she would have the powers of a Senator. If the President's party controlled the Senate the Vice President would be the leader of the majority party, otherwise he or she would be the leader of the minority party. Promoting the President's legislative program would provide the Vice President excellent training in the event he or she were called upon to assume the Presidency.

Second, given the national election of senators and their control over legislation, the annual election of senators becomes a referendum of the President's legislative program. To promote his program, the President would have to campaign vigorously for his party's candidates. Failure to capture a majority of the newly elected senators would be taken as a signal that the electorate wants compromises between the positions of the parties. Also, a President would become a lame duck if his party lost too many senatorial elections.

Now let us consider the role of the House in the proposed specialization of legislative functions. The House would focus on the day-to-day functioning of government. The main functions of congressmen, a traditional term assumed to include congresswomen, would be to resolve governmental problems of their constituents, to conduct administrative oversight, and to vote on sponsored legislation. Unlike today, constituents would only approach their congressmen with governmental administrative problems. Congressmen would have a technical support staff for administrative oversight and a larger staff to handle constituent relations. Congressmen who maintained good constituent relations, provided oversight on administrative matters of interest to their districts, and voted the interests of their districts could enjoy a long career. Even though all their actions would be subject to a possible professional review, congressmen would enjoy less risky careers than those of senators.

The power of the House to influence government legislation and policies would stem from its power of legislative oversight. The House alone would investigate scandals and other failures of leadership in the administration. Also in keeping with tradition, the House would initiate the legislative budget process. This process would provide congressmen with clout in their oversight in how well administration policies were satisfying their constituents needs. Although congressmen could neither sponsor nor amend legislation directly, they would have considerable indirect influence on the content of legislation because they would have to approve all sponsored legislation. In proposing sponsored legislation the Senate would seek input from the House.

The new design creates greater checks than currently between the House and Senate, especially when the majority of the two houses are from the same party. The interest of senators in ensuring that their proposed legislation will pass muster in a professional review places them in conflict with congressmen accountable to local interests.

Executive

The President in informational society would have much the same role as in the current government. With a constantly changing political economy, the President needs much greater power to reorganize the bureaucracy according to changing conditions. Additionally, to achieve a more flexible government, the President needs authorization to replace much of the permanent civil service with the management situations market. However, giving the President more control over the administration, requires a reconsideration of the system of checks and balances. Executive action would be subject to a professional review and the court and Congress would have expanded oversight powers. The interaction of the proposed Congress and the President would create incentives for much greater efficiency in government.

The administration of government, like that of business and other institutions will change continually as knowledge and technology advance. In time, the flow of information between various agents in society and the administration will primarily take place through one machine communicating with another. In particular, much of the routine interaction between the private sector and government will be automated. For instance, a citizen requiring a license or needing to enact other routine business with a government agency will communicate with an interactive program. Automation in government administration, like automation in business will thereby displace routine jobs, leaving the work to be increasingly performed as a succession of one time jobs.

Given the environment of constant change, the administration, like corporations, requires frequent reorganizations to best fulfill its functions. In the current government design, reorganization is a very difficult undertaking which requires legislative approval. It is hampered by the fact that legislators in committees and subcommittees guard their turf and generally resist attempts to reorganize. Consequently, a fundamental problem with the current government bureaucracy is inflexibility. With changing needs and technology, the President needs much greater authority to reorganize the bureaucracy to meet these changing conditions. In the proposed design the President would be granted the power to reorganize the administration annually. The reorganization plan would be submitted simultaneously with the submission of the budget. If two thirds of Congress should veto the reorganization plan within a month, the plan could not be implemented.

The purpose of this executive power would be to enable the President to adjust the organization of the administration to accommodate changing conditions. Currently a variety of agencies are responsible for regulating financial institutions. However, because changing laws have now removed the legal differences among most aspects of financial institutions, it would now be wise for the President to move all the elements of regulation under one agency. With the power of reorganization the President would be the agent responsible for obtaining an effective organization. A flexible administration enables the President to create temporary organizations to handle critical current problems. With such power to reorganize government, for instance, President Reagan might have created a high-level organization to battle drugs. When some future President felt the drug problem had been overcome, he could then revise the organization.

Due to constant changes in government the concept of lifetime employment implicit in the civil service would be replaced with finite-time contracts obtained through a management situations market. These finite-time contracts would enable the President to address the changing succession of one-time governmental problems. Under such conditions the role of the civil service commission would be to ensure that competition for positions was based on merit and not politics. One approach to this promotion of merit might be to place the decision process for the administrative management situations market in the public domain. The political-appointment-decision process, however would be conducted in private, since these appointments would be reviewed by Congress.

The power of the President to reorganize the administration would give the President the power to search for effective organizations for changing conditions and technology. Because a bounded rational manager can only manage a small number of subordinates, the President would probably organize his cabinet with only five to seven cabinet officers. And each cabinet officer would have a small number of subordinates and so on. Besides appointments in the hierarchy, the President would make a small number of appointments to independent agencies. In reorganizations Presidents would shift responsibilities from one cabinet officer to another. For example, one President might wish all activities promoting science and technology under a single science official, while another President might wish these activities split up among various departments. If the President wished to place all civilian and military purchasing under an efficiency expert, he could do so. Also, legislation authorizing the President to perform some new governmental activity would simply authorize the activity then allow the President to organize the activity in his administration.

Granting the President new powers in addition to the powers acquired by the President in the second hundred years raises the question of appropriate checks and balances. As was pointed out, Congress, to ease its legislative burden, must delegate considerable policy details to the President in administrating legislation. Because this delegation of legislative details to the executive was not the practice in 1780s, the legislature was not provided any specific checks on how the executive implements legislation. In practice, Congress upholds a nebulous concept called legislative oversight. While Congress has a legitimate right to investigate the administration for the purpose of modifying the law, Congress impinges on the separation of powers doctrine when Congress investigates the administration in order to manipulate legitimate executive policy.

The current status of constitutional interpretation is that the legislature does not have the right of a legislative veto over individual executive actions unless the veto is submitted to the President as a piece of legislation[12]. Congress must resort to political investigations to force resignations from officials who seriously deviate from legislative intent. Congress needs some mechanisms of legislative oversight which would check, but not incapacitate executive capacity to organize and act. Congress should have the right to specify the conditions under which the President can fire an administrative official in charge of an authorized government activity, the right to specify the information policy of the authorized activity, and the right to recall a political appointee in the executive branch.

These rights would give Congress some influence over how the President organizes and conducts his administration. In every piece of legislation authorizing executive activity the legislation would specify the conditions under which the President could fire the official in charge of the activity and the information policy the activity would have to follow. In executive activities traditionally controlled by Presidential policy, the President would have the right to fire political appointees at will and the information policy might have a time delay release for politically sensitive information, such as the conduct of foreign affairs. In executive activities which Congress felt should be independent, Congress might specify that the President could fire political appointees for cause and would have to operate with full public disclosure. In some government activities Congress might go so far as to require that all external communication, such as with other government agencies, the White House staff, and private concerns, be part of the public record. Congress might explicitly limit the concept of executive privilege to an inner circle on the White House staff. To be efficient in organizing the administration, the President would probably organize activities with similar firing and information policies together.

In addition, the proposed right to recall appointments simplifies the legislature's problem in obtaining compliance with legislative intent. The current procedure for obtaining compliance with legislative intent is to embarrass the President with legislative investigations until the President asks the official to resign. With the right to recall Congress could simply recall the political appointment, provided the official were a member of the executive branch. If two thirds of congress votes for a recall, recall is automatic and requires no justification, however Congress would undoubtedly hold a media oriented investigation to justify their action to the public. This provision means that a President for whom the legislative majority is from the opposing party must in his policies maintain the support of one-third of Congress. The effect of the recall, then, would be to force the executive to compromise with the legislature when the President's party was a minority in Congress. The resulting compromises should achieve greater consistency between legislative intent and policy implementation.

In the proposed design, in addition to legislative checks, all executive action would be subject to a professional review by the courts. If a party were affected by executive action, he could, as was the case with legislation, file a professional review case based on professionally refereed analysis in federal court. Subjecting executive action to a professional review would nullify the concept of the Administrative Procedures Act which limits judicial review to legal issues such as due process. Besides desirable legal properties such as due process, executive action, that is all administrative actions, should have the properties of consistency, general benefits, and efficiency. The professional review of executive action would influence both the organization and the implementation of legislation. As a President would have to carefully consider his organizational plan, professional challenges would very likely be few. Professional challenges of executive action would be much more likely in government activities in which the legislature granted the executive broad powers in implementing the legislation.

Finally, the courts would have one new check on executive action. Frequently private citizens sue the government to force administration officials to carry out the provisions of legislation. Such activities can occur in a change of administration where the new administration does not like the legislation of previous administrations and decides to change the law by a policy of benign neglect. If the plaintiff wins and the court decides the failure to follow the law was willful, the court as a deterrent should have the power to remove the responsible officials from office. The purpose of this power is to encourage administration officials who do not like particular laws to submit the desired change in the next presidential legislative program rather than change the law by policy implementation.

In response to greater judicial checks, the administration would take greater care to ensure that policy complied with legislation. Moreover, to greatly reduce the prospect of negative professional reviews, the executive would establish its own executive professional review. As both executive amplification of legislation and budget items are subject to a possible professional review by the courts, the executive would create administrative professional review procedures to guard against subsequent court challenges. Cautious bureaucrats would ensure that administrative action, proposed legislation, and new budget initiatives were carefully reviewed by professional proponents of all major theories. Most professional reviews would include a professional teleconference which pitted the analysis of the various positions against each other with a panel to judge the proceedings.

The proposed modifications to Congress and the Presidency would also create incentives for much greater efficiency in federal government operations. For example, greater efficiency would result from changes in the budget process. In order to regain some of the control it had lost to the President, Congress would insist on shifting the budget process from a line-item budget process to an output-oriented budget. Line item budgets define the budget in terms of inputs used by government. In a static administration, legislative committees, through experience, intuitively know the relationship between inputs and outputs[13]. Through cozy relationships with the permanent bureaucracy, legislative committees can exercise oversight to provide services that constituents want, for example, a road in a national forest to harvest timber. To strengthen his control over the administration, a President, then, has incentives for reorganizing to break these ties. To regain some measure of this control Congress would have incentives to switch from a line-item budget process to a budget process such as zerobased budgeting which relates inputs to outputs. To the extent that the budget could be precisely defined in terms of the relationship between inputs and outputs, Congress could control the level of administrative outputs independent of Presidential reorganizations.

With more quantification made possible through more computerization, budgets relating inputs to outputs could be implemented; however, the development of such a form of budgeting would be a monumental task. For government services controlled by software the relationship between inputs and outputs is clearly defined in the operation of the software. The relationship between inputs and outputs in activities such as power generation is as well known in public generation as it is in private generation. Some government activities, such as the activities of the Attorney General, for which the output is at best fuzzily defined, would resist quantification for a long time. With a zero-based budget process, the focus of the budget process will be on the desired levels of output. Computer programs using the proposed output levels would then generate the input expenditures. For entitlement programs, then, the forecasted level of service would determine the budget.

Without greater incentives to promote efficiency, the move to a form of budgeting relating inputs to outputs alone is insufficient in itself to promote efficiency. The key which would create incentives for greater efficiency would be the proposal that each budget item would have to be sponsored by a nationally elected senator who would then have incentives to analyze the efficiency of budgeted items in order to avoid a professional review. This contrasts sharply with the current situation under geographic representation where constituents in each district primarily regard their representative as a procurer of government services for the district. As efficiency in providing these services would generally have a negligible impact on taxes, efficiency currently becomes an abstraction relating to government in general.

Nationally elected senators constrained by the professional review, however, would not be able seek voter approval by trying to provide specific services to narrow constituencies. They would have to seek national constituencies. Consider the problem of the senators on the committee for sponsoring the budget. With a small national elected budget committee, the incentives for efficiency would have been greatly modified. The budget committee focuses such national concerns as efficiency, the level of spending, and the deficit into a small group of nationally elected senators. The successful budget senator would want to increase government services to please constituents and at the same time reduce the level of spending in order to reduce taxes.

Such an approach would seem impossible until one considers the subject of government efficiency. With the advance of technology, government operations, like private operations, are capable of great increases in efficiency[14]. The senate budget committee would create incentives for congressmen to concern themselves with efficiency of activities for which they had oversight responsibilities. In the bargaining process to obtain votes for the budget, the senators would trade improvements in efficiency for increases in activities of interest to constituents of congressmen, provided, of course that they were within the constraints imposed by a professional review. A successful budget senator could then campaign on the basis of how he increased services through efficiency and even achieved a slight decrease in taxes. A successful congressman would be able to claim an increase in the type of service desired by his constituents.

Furthermore, congressmen in the administration oversight committees would obtain increases in administrative efficiency through a variety of approaches. One of these would be to authorize research to define more clearly the relationship between inputs and outputs. For those government services where these relationships could be defined, incentives systems could be created to promote productivity increases. Inevitably, part of budget outlays would be hardware and software enhancements to achieve productivity advances. Where relationships between inputs and outputs could be defined, improved performance by government managers could be rewarded by financial incentives. Part of this improvement would occur through the replacement of the permanent static bureaucracy by privatization[15] and management situations markets.

Evaluation

The key to good government is to create the appropriate incentives to use knowledge to translate vaguely defined general purposes into specific objectives and policies. In representative government politicians compete for power to articulate these objectives and policies and the general voter holds elected politicians accountable for their actions. The professional review improves the incentives in the system of checks and balances by providing the general voter with an easily understood measure of political performance which voters would use to evaluate incumbents. Nevertheless, the professional review would have strong incentive effects on the behavior of elected and appointed officials only if such officials assumed that a professional review case would be filed the minute policy research established a government defect.

Since all groups in society are negatively impacted by bad legislation and administration, any individual, firm, public or private institution could file a professional review case in court, but generally few would file such cases unless the estimated gain was greater than the estimated costs. Whether the professional review creates incentives for elected and appointed officials to incorporate general benefits, consistency and efficiency in all their actions depends on whether private parties would have sufficient individual incentives to file a professional review case on governmental defects established by analysis.

Pursuing a professional review case would be costly. First, the court to avoid being overwhelmed by nuisance suits would require considerable impartial evidence such as professionally reviewed research to initiate the case. Second, pursuing a case to the Supreme Court is very expensive.

Finally, there is a free rider problem in that, while the instigators of the case would have to pay their legal fees, the benefits are, frequently, widely distributed, and the instigators have no mechanism to recover costs from the beneficiaries. Thus, it is not obvious that sufficient professional review cases would be filed to create an incentive effect on legislators and administrators.

From the perspective of costs, two factors lesson the burden. First, from the perspective of prospective initiators of professional review cases, the professional review is simply another tool that might be used to influence government. Although pursuing a professional review case would indeed be expensive, it would be much cheaper than trying to mount a public relations campaign to galvanise public opinion into demanding legislative action. Second, as will be discussed in the next chapter, the right to learn could be interpreted as a requirement that the government must fund professional-review research, and to be efficient, such research monies would have to be awarded by an impartial peer group award system. This means that much of the research exposing conditions warranting a professional review will be publicly funded. Private sources would fund additional professional review research which would be performed by think tanks such as the Brookings Institute and the American Enterprise Institute. Once research has been published in a reputable journal, any party could use this research to initiate a professional review case.

The motivation of firms in filing professional review cases would be self-interest. Consider the implications of a professional review on the current political economy. Previously the inconsistency in risk policy between old risks and new risks was discussed. Trade associations for firms subject to the more costly, higher standard screening for new risks would have a strong incentive to file a professional review case to receive treatment consistent as with that of old risks. The possible result of such a case might be to force the administration to implement a policy of comparative risk between old and new technology. Trade associations would have incentives to challenge regulations such as the environmental regulations which require uniform reduction in pollutants by all polluters. Most economists would argue that general benefits would be increased with a system which reached targets but allowed a market mechanism to determine the distribution of pollution between firms.

On occasion, a trade associations might pursue an efficiency case to compel the administration to adopt technology proved efficient in similar operations in private industry. Many government activities, for instances, the processing of information have similar operations in business. If the government falls behind accepted business practice, the vendors of the new technology would have incentives to press efficiency suits. Moreover, on rare occasions the trade association of one industry might use a professional review to challenge a government granted privilege, for example preferential tax treatment, to a rival industry.

Besides purely private professional review cases public agencies and public interest groups would file professional review cases to achieve their goals. A new President, who represented a shift in political philosophy from his predecessors, would consider the professional review a potential tool to prune the legislative legacy of his predecessors. Also, lower level governments would on occasion use the professional review to challenge legislation and executive actions of higher level governments.

Public interest groups such as the Sierra Club, Common Cause, and Ralph Nader's group would use the professional review to promote their causes. They would probably ask that tobacco subsidies be eliminated as both inconsistent and lacking general benefits. Public interest groups would also insist on the basis of consistency that the oil industry be included in the effort to clean up toxic wastes with the Superfund. In addition, Mothers Against Drunk Driving, MADD, would probably press for a more consistent treatment of drugs and alcohol. Public interests groups would challenge the practice of bailing out failed large banks which protects the stockholders' assets and selling off the assets of failed small banks which generally leaves the stockholders with no assets. The small banks trade association might pursue this in the remote hope of obtaining a similar subsidy; however such a subsidy would be challenged as lacking general benefits.

Many general benefit and efficiency professional reviews would be initiated by citizen watchdog groups such as taxpayers associations. Under general benefits, those activities which lack national benefits should be funded locally or by the recipients. For example, watchdog groups would use a professional review of general benefits to halt the efforts of the federal government in trying to protect homes built on barrier islands from the Atlantic ocean. Such groups would use efficiency professional reviews to obtain faster enactment of efficiency studies. President Reagan created the Grace Commission to study the efficiency of the federal government. After some study, the Grace Commission proposed that through greater efficiency the federal budget could be reduced some 400 billion dollars without reducing services. Many of the Grace commission recommendations will be enacted over time, but re-election conflicts, such as closing military bases, means the process will be a slow one. The professional review challenge of efficiency would speed up the process of achieving efficiency in government.

Since many groups would have an incentive for pursuing judicial professional reviews, the number of these reviews should be sufficient to create the perception that, whenever research revealed deviations in legislation and administrative actions from the precedents set by professional reviews, some interested party would surely file a judicial professional review case. If this were the case, senators and administrators would carefully consider the precedents established in professional reviews in proposing legislation and administrative actions. Moreover, senators and administrators would take steps to correct any defect revealed by research prior to a trial. Once this happened, the number of actual judicial professional review cases to maintain the incentives could be quite small. The main effect of the judicial professional review would be the incentive effect on Congress and the administration.

One consequence of the incentive effect of the professional review would be that given publicly funded elections, the relationships between senators and lobbying groups would change. Senators, while not beholden to lobbying groups for campaign finance, would still need to maintain support of interest groups in a broad coalition in order to be re-elected. A senator would therefore be interested in legislative proposals of interest groups in his coalition. In providing the type of legislation wanted by his coalition a senator would have to take precautions to reduce the possibility of a rash of negative professional journal articles, or even worse, a successful negative professional review. Senators would insist that they had a budget for contract research to carefully analyse legislative proposals. They would also have an incentive to have each legislative proposal reviewed by prospective opponents of the proposal in order to ferret out those proposals which would not stand up to a professional review. A successful lobbyist would therefore need to be able to create proposals which could survive a potential professional review and, in attacking legislative proposals, know how to create a reputable analytic argument. A lobbyist's status would thus be based on both his persuasive and his analytic skills. In this way the professional review and public funding of campaigns does not eliminate lobbying, it merely changes the incentives.

A second consequence of the incentive effect of the professional review would be research to continually refine the precedents for establishing general benefits, consistency and efficiency. Liberals wishing to promote government expansion would promote research to establish new forms of measurement of the benefits of government. At the same time, conservatives would promote research to demonstrate defects in positive measures of government benefits. Regardless of the political motivation such research would constantly refine the precedents for establishing general benefits. Likewise, research would constantly refine the precedents for consistency and governmental efficiency.

And a third consequence of the incentive effect of the professional review would be a better adjustment to the accumulation of knowledge. Improvements in the estimation of the common weal are to be achieved by increasing the use of knowledge in this endeavor. As the social sciences and related disciplines advanced a basic requirement of general benefits would be established that legislation and administration be based on a complete formal analysis. Even with competing theories of behavior, the basic requirement for a thorough analysis would result in better government, since complete analysis places a check on poorly conceived legislation. If the analysis were so superficial that it is incorrect with respect to the stipulated theories, the law or administrative action would quickly be declared unconstitutional by a professional review. If the theories upon which the analysis supporting legislation were correct, then the law and administrative actions based on careful analysis would be superior to those based on *ad hoc* intuition. In the case where the theories upon which the legislation was based are subsequently proven incorrect, the professional review would provide a mechanism for much more rapid removal of the legislation than obtaining a majority vote of the legislature. As soon as most of a profession rejected the theories upon which a law are based as false, the initiation of a professional review would result in the law being declared unconstitutional. As knowledge advanced the performance of legislation based on complete analysis should increase over the current practice of at best partial analysis.

The estimation of the common weal can also be improved by innovations in the strategy for implementing governmental innovations, a topic which will be considered in the next chapter.

Notes and References 8

1. This effort is in the spirit of the founding fathers in asking the question: What is an appropriate system of checks and balances for the 21st century. In searching for references for this book I was delighted to discover that the social choice theorists, after demolishing any prospect for a fair voting system,

place great emphasis on the system of checks and balances. For example see: Riker, W. H., 1982, *Liberalism versus Populism: a Confrontation Between the Theory of Democracy and the Theory of Social Choice*, (W. H. Freeman & Company: San Francisco). Buchanan and Tullock provide a social choice analysis of constitutional democracy in their classic: Buchanan, James M. and Tullock Gordon, 1962, *The Calculus of Consent: Logical Foundation of Constitutional Democracy*, (University of Michigan Press: Ann Arbor)

2. This approach is pursued in Masuda, Yoneji, 1981, *The Information Society as Post-Industrial Society* (Institute for the Information Society: Tokyo)

3. In the 1960s Arthur Kantrowitz proposed a science court for making scientific judgments. For example, see: Kantrowitz, Arthur, 1967, Proposal for an Institution for Scientific Judgment, *Science*, 12 May, pp 763-764. This proposal was vigorously debated. For a negative analysis see Sofaer, Abraham, 1978, The Science Court: Unscientific and Unsound, *Environmental Law*, Vol 9:1, pp 1-27. One of Professor Sofaer's arguments is that the administration has the technical competence to make such judgments and the science court is unnecessary. In this design the Administrative Procedures Act is to be repealed and the Courts must make professional judgments. The judges in fact are necessary to attain separation of powers for judgments in fact as well as judgments in law.

4. Recently (1991) Peter W. Huber made a passionate case in *Galileo's Revenge: Junk Science in the Court Room* (Basic Books: USA) that economic incentives, a relaxation of the Frey rule and the influence of Calabresi replacing common law has lead lawyers in liability cases to base their cases on the expert testimony of disreputable expert witnesses who, for a fee, will provide bogus plausible causation to win large settlements. Peter W. Huber recommends judges(-in-law) impose the Frey rule which would require expert witnesses to be mainstream scientists. In the design of this book, judges-in-fact would determine both the competency of expert witnesses and the standards of expert evidence. Over time through precedents, standards for expert testimony would be established by these judges-in-fact.

5. The legal profession is well aware of the shortcomings of the advocacy trial. For example, see: Frankel, Marvin, 1976, From Private Fights Toward Public Justice, 51 N.Y.U.L. Rev, 516. It might appear to the legal profession that adopting cheaper dispute settlement procedures will lower their revenue. If the demand for legal services is elastic, however, creating less expensive dispute settlement procedures would enable lawyers to market their services to the masses, not just the wealthy, and as a consequence increase not decrease their revenues.

6. The movement to analyze the impact of law on society goes back at least as far as the late 19th century. For example, see Monahan, J. and L. Walker, 1985, *Social Science in Law: Cases and Materials*, (The foundation Press, Inc.: Mineola, NY). The requirement here is that this analysis should consider the social impact from the perspective of all relevant disciplines. Professor Ackerman argues that the law needs to be analyzed on a broader basis than the Realist movement's concern for particulars and avoidance of formal analysis.

This is already taking place in tort law with the intrusion of Coase's theorem. See Ackerman, B. A., 1984, *Reconstructing American Law*, (Harvard University Press: Cambridge). The creation of judges in fact would accelerate the trend towards formal analysis of law and legal precedents by the relevant disciplines.

7. From 1897 to 1937 the Supreme Court used the economic interpretation of the 14th amendment to strike down state government attempts to regulate industry as an intrusion in the right of contract. The current criterion is much weaker than the economic interpretation of the 14th amendment. If almost all professionals believed that for a particular regulatory activity, the cost of government regulation was greater than the social benefits, the new criterion would achieve the same result as the economic interpretation of the 14th amendment. If professionals were divided as to the costs and benefits of government regulation, the regulation would pass the new criterion.

8. See: Huber, Peter, 1984, Discarding the Double Standard in Risk Regulation, *Technology Review*, Jan, pp 10-14

9. With the advancing social nervous system the possibility of implementing revealed demand mechanisms increases. For example, see: Green, J. and J. Laffont, 1979, *Incentives in Public Decision-Making*, (North Holland: Amsterdam)

10. In a clinical study of several recent political innovations Polsby discovered that the more controversial an act the more carefully it tended to be analyzed. See: Polsby, N.W., 1984, *Political Innovation in America: The politics of policy initiation*, (Yale University Press: New Haven)

11. I assume the candidates would be elected by simple majority vote. In the event some candidate failed to obtain a majority the voting procedure could be either alternative vote or perhaps a run-off of the two candidates with the most votes. Social choice theorists have demonstrated the negative conclusion that all voting systems are liable to manipulation. See: Gibbard, A., 1973, Manipulation of Voting Schemes: a General Result, *Econometrica*, 41, pp587-601 and Satterthwaite, M. A., 1975, Strategy-Proofness and Arrow's Conditions: Existence and Corresponding Theorems for Voting Procedures and Social Welfare Functions, *Journal of Economic Theory*, 10,pp187-217

12. *Immigration and Naturalization Service v. Chadha*, 103 U.S. 2764 (1983)

13. The first input-output budget initiated at the federal level was planning-programming-budgeting(PPS) in 1965 at the Pentagon under Robert S. McNamara. This form of budgeting was extended to the entire government under the Johnson administration but then dropped by the Nixon administration in 1971. Another form known as zero based budgeting(ZBB) was tried without great success by the Carter administration. This form of budgeting upset political relationships, required many more analysts than existed and underestimated the conceptual problems in determining the relationships between inputs and outputs. For a discussion of this form of budgeting see, for example: Lyden, F.J. and E. G. Miller (Eds), 1982, Public Budgeting: *Program Planning and Implementation, 4th edition*, (Prentice-Hall, Inc: Englewood Cliffs, NJ). This

form of budgeting has severe critics. For example, see: Wildavsky, A., 1988, The New Politics of the Budgetary Process, (Scott, Foresman and Company: Glenview, Ill). With advances in economics and operations research over time the number of analytic tools to analyze the relationships between inputs and outputs will increase. Very gradually budgets based solely on inputs will be replaced.

14. Ronald Reagan appointed J. Peter Grace to head the Private Sector Survey on Cost Control. The Grace Commission discovered that $424 billion could be cut out of the budget by eliminating waste and efficiency. For a summary of the work of the Grace Commission see: Kennedy, W. R. Jr and R. W. Lee, 1984, *A Taxpayer Survey of the Grace Commission Report*, (Green Hill Publishers: Ottawa, Ill). To achieve efficiencies such as suggested by the Grace Commission would require much stronger incentives for efficiency than are currently the case.

15. See: Savas, E. S., 1987, *Privatization: The Key to Better Government*, (Chatham House Publishers, Inc: Chatham NJ)

Chapter 9

Governmental Decentralization

Introduction

In this Chapter additional governmental modifications to promote governmental innovation will be proposed. As was pointed out in Chapter 1, the rate of governmental innovation should be commensurate with the rate of private invention and innovation. And as was pointed out in Chapter 2, the current rate of governmental innovation is seriously deficient. Modifications to government were proposed in Chapter 8 to promote governmental innovation through a better estimate of the common weal. In this chapter modifications to obtain a better strategy for implementing governmental innovations will be proposed.

To promote a level of public innovation commensurate with private invention and innovation, the government would greatly increase its promotion of two types of discoveries. The first type would be discoveries which create new opportunities for governmental innovations. Discoveries in many disciplines would promote governmental innovation. For example, the future governmental control of negative externalities such as pollution would place much more emphasis on creating incentives for the participants to reduce the negative externality themselves. Consequently, discoveries in the social sciences which would provide new insights into the implications of alternative incentives systems would create new opportunities for innovations in environmental incentive systems. For another example, some discoveries in operations research would create new opportunities for governmental efficiency. Also, some types of computer science discoveries would create innovations in governmental data processing.

The second type would be discoveries which provide new improved tools for analyzing alternatives in governmental tasks. A governmental innovator, like other innovators, must forecast the potential performance all the proposed alternatives. The long range goal should be to make the analysis of potential alternatives for innovation much like the current analysis of alternatives in invention using tools of computer assisted engineering or like the analysis of business alternatives using spreadsheets. For governmental innovation, the software tools for analyzing the alternatives would be simulation programs to forecast the cost and behavioral consequences of selecting each of the alternatives. Consequently, in order to improve innovation performance, government should take steps to promote discoveries which would increase the predictive capability of the social sciences.

To promote these two types of discoveries, governmental modifications would include revising current information policy and much greater governmental research support for these two types of discoveries. The scientific information policy discussed in Chapter 7 would increase the rate of discovery promoting public innovation. In addition, as was proposed for private innovation, government would increase the funding for basic research promoting public innovation, and part of this funding increase would sponsor a much larger number of social experiments. Such increased innovation-promoting research funding would be required to better balance the high levels of invention-promoting research funding.

However, many more governmental modifications to promote innovation are required in addition to creating better tools for analyzing alternatives and new opportunities for innovation. Even with the professional review to keep estimates of the common weal in compliance with the precedents established for general benefits, consistency and efficiency there will generally be competing estimates of the desirable social objectives for any task assumed by government. For example, for the foreseeable future there is unlikely to be any precise agreement on the best tradeoff in the conflict between environmental preservation and economic growth. Empirical observation of the consequences of different objectives would be necessary in order to obtain a better estimate of the common weal.

And in the foreseeable future, while the predictive performance of the social sciences is likely to improve considerably, the social sciences are unlikely to develop to the extent that the consequences of government alternatives can be precisely forecasted prior to the actual implementation. This means that simulation tools will be useful for reducing a large number of potential alternatives to a few promising alternatives, but that actual implementation of each promising alternatives would be necessary to accurately rank them.

Finally because of the limits of knowledge, obtaining good performance out of the implementation of a chosen policy alternative would require much applied discovery in order to achieve the best results from that alternative. A critical factor in governmental innovation, then, is the strategy for implementation.

Given such difficulties, the key to developing a better strategy for implementing governmental innovations is the creation of a more rational criteria for governmental decentralization than the current purely political criteria. To consider how this might be accomplished let us start by reviewing the current status of governmental decentralization. The Supreme Court interpretation of the 10th amendment determines the amount of decentralization from the federal government to the states. Originally the states were to have all powers not explicitly assigned to the federal government. With the growth of the federal government in the 20th century, the Supreme Court creatively interpreted the 10th amendment to sanction the expanding role of federal government. Under the Garcia precedent, federal decentralization to the states is now solely determined by the federal political process. And since the 19th century Dillon rule, government decentralization below the level of the states is totally a prerogative of the state governments[1]. Thus current decentralization of governmental activities is determined by legislative whims at the federal and state levels. For example, the Bush administration in order to reduce the federal deficit is decentralizing federal governmental activities to the states.

What would curtail legislative whims in governmental decentralization in informational society is the professional review. Under such a review governmental decentralization, like all other governmental acts, would have to have the properties of general benefits, consistency and efficiency. By the definition of governmental innovation if implementation of a new governmental decentralization criterion leads to a higher rate of innovation, it implies better performance as measured by the true, but unknown common weal. Thus such a new criterion would have general benefits. The consistency criterion implies that governmental decentralization should operate under a consistent policy which would be created by the development of rational criteria for decentralization.

The efficiency criterion implies that government activities should be assigned to the level of government at which they would be most efficiently performed.

In this chapter we shall propose rational decentralization criteria which balance the need to place governmental activities at the most efficient level against the need for greater variation to achieve a higher rate of innovation. Also in order to obtain more effective strategies for implementing governmental innovations, we shall propose that the current government be restructured to place governmental activities at their proper level. Once this is done the professional review would be employed to maintain rational criteria for governmental decentralization which in turn would result in improved strategies for implementing governmental innovation. We shall then forecast the operation of the decentralized government of informational society.

Ideal Governmental Decentralization

The starting point for reorganizing governmental decentralization in order to increase governmental innovation is to consider the defects in the current governmental decentralization. The original design attempted to completely separate the functions of federal and state governments to achieve a concept of dual sovereignty. However, in the subsequent expansion of government most new government functions were decentralized with more than one level of government performing tasks of the new functions. For example in the implementation of the Clean Air Act, the federal government set the standards and states and cities had to submit plans for implementing the standards.

The problem is that the current decentralization of governmental tasks does not pay much attention to either efficiency or effective learning. An example of inefficiency is that funding social programs at the federal level and operating them at the state level leads to overpayments[2]. An example of failure to learn is that in developing a plan to implement the original federal clean air standards, the cities and states were not allowed to experiment with alternative incentive systems to achieve the targets[3]. This lack of variation seriously limited empirical learning needed for improving performance over time.

A rational criteria for the decentralization of governmental tasks must balance considerations of improving governmental innovation with considerations of governmental efficiency. Let us first consider how appropriate decentralization of governmental tasks can improve innovation. Currently, the most common strategy for implementing governmental innovations is the improvisatory strategy where government implements a plan and improves the performance through learning by doing. Decentralization of government tasks would greatly improve strategies for implementing government innovation as decentralization would enable government to shift from an improvise to a separation strategy for innovation. As was pointed out, the separation strategy, which is applicable to cases with a large number of similar tasks, is implemented by separating the research task from the operational task.

With a separation strategy for governmental innovation the governmental function is partitioned into tasks with the federal government performing research and development while subordinate governments perform the operational tasks of the government function. This strategy has been used since 1900 in agricultural research and since since World War II has been initiated in some types of social policy. National social policy now is increasingly based on pilot projects, demonstration studies and social experiments. For example, the social experiments in income maintenance provide some empirical knowledge for future tax policy for low income groups.

Nevertheless, because most research promoting governmental innovation would involve humans, such research would face restrictions not found in research on inanimate objects. This is because in many cases systematic experimentation requires negative treatments on some individuals. For example, positive variations in copayments with government medical insurance to test alternative demands for medicine can be construed as a negative stimulus, since a destitute person would conceivably forego a treatment which could save his life. In informational society as under current law[4] variations which could have a negative effect would require informed consent. Thus an experimental design which wished to incorporate negative or risky treatments would have to consider the need to compensate subjects for accepting the risk. Consequently where possible, research in governmental functions would focus on variations in treatments where all the variations have a neutral or positive effect on the participants. Such would be the case in the testing of new educational software on selected schools with a control group which does not have the new software.

However, as the governmental research task would be performed regardless of whether the operational tasks of the governmental function were decentralized, the advantage of decentralization of operational tasks requires further clarification. In the proposed governmental decentralization if a task is assigned to a subordinate government, this government must be granted some control over both the setting the objectives of the task and selecting the alternatives to achieve the selected objective. And as will be subsequently discussed the subordinate government could file for an experimental variance to gain greater control than granted by the superior government. For example in the current decentralization of medicaid to the states, the states have little control over the objectives of the program or the choice of alternatives. Under the proposed type of decentralization if the task of providing medical services were decentralized to the states the states would determine both the objectives and the policy to achieve the selected objective. Thus with the decentralization of a task, it is likely that different subordinate governments will select different objectives as being the best estimate of their respective common weal. Such empirical differences are needed to improve the estimate of the common weal over time.

In addition, with the decentralization of a task the differences between subordinate governments in terms of political beliefs and accepted theories also make it likely that more than one policy alternative would be implement by these subordinate governments. Decentralization would also increase because policy variations among subordinate governments would not require informed consent. For example, if operational control of medical insurance were passed from the federal government to the states, variation in copayments among the states would not require informed consent of the recipients of medical insurance unless a state wanted to vary copayments within the state. Thus decentralization of governmental tasks is very much likely to increase the variation in risky tradeoffs. To further enhance variation on a more systematic fashion, higher levels of government could offer lower levels of government incentives to try new alternatives. Finally, with more governments implementing alternatives competition will encourage immediate implementation of any successful features in implementation.

An additional factor favors decentralization in promoting innovation. The federal government's role in research would be enhanced because part of the research role would be in carefully measuring and evaluating the differences in objectives, policy alternatives and their implementation by subordinate governments. This analysis, as will be subsequently demonstrated, would be distributed to voters is such a fashion that it would galvanize them into demanding higher performance from their elected officials in the subordinate governments. Such demands would increase the properties of general benefits, consistency and efficiency in the estimates of the common weal by subordinate governments. And the competition between subordinate governments would greatly increase the concern for efficiency leading to much more rapid imitation of successes. Also, decentralizing activities from the federal government would simplify the federal government which consequently, would reduce the amount of intensity bias at the federal level.

Now let us consider the efficiency factor in governmental decentralization. It is obviously inefficient to decentralize some governmental functions such the military, foreign policy and monetary policy. Although such governmental functions should not be decentralized, the government should consider more systematic policy variations over time to learn more; however, this topic will not be pursued in this book. Decentralization of other governmental functions, such as automobile safety standards, could add costs to national businesses who would have to modify their products to meet subordinate government policies. Such businesses could lose economies of scale in production. Thus, the efficiency of decentralizing a governmental task must consider both the efficiency of performing the task itself and its impact on the political economy.

The development of a decentralization plan must be based on balancing efficiency and effective learning. For this purpose, government should be more decentralized and have greater variation at lower levels of government than is the case today. Most groups in society seriously underestimate the need for variation in government. Frequently in public policy debates, such as national educational testing or workplace child care, one faction or another raises a clarion call for the need for a national policy. Such groups fail to understand that a national policy is a poorly defined experiment with a sample of one. Given equal treatment under the law, nothing is learned about the alternatives. In as much as social problems usually continue for decades, variation is needed to obtain empirical information in order to make systematic improvements in performance over time.

In addition, advances in the political economy are making increasing decentralization more efficient. In the 1930s there were economies of scale in placing social programs at the federal level because of the small number of trained experts in the political economy. Today with the great expansion of professional education, the supply of trained experts is sufficient to enable all levels of government to obtain competent help. Because the need to place governmental tasks at high levels of government in order to obtain experts is not likely to be a factor in the future, administration of smaller programs can be just as effective as larger programs. In addition, advances in automation and the resulting reduction in the costs of batch production will enable industry to respond much more cheaply to variations in regulations which are needed to promote learning. Finally, advances in decision support systems will enable decision makers to quickly, cheaply deal with greater variation in governmental policies.

To implement this decentralization theory into government, two steps are proposed. First, the various functions and tasks of government would be initially reorganized in order that a separation strategy for governmental innovation could be employed whenever decentralization is efficient. Second, decentralization criteria would become part of the professional review in order to maintain ideal decentralization as the political economy advances.

Thus under the proposed reorganization, operational control over current governmental functions and tasks would be transferred to the lowest governmental level where they can be efficiently performed. And for this criterion efficiently refers to the overall political economy and not just government. The federal government would fund most research promoting innovations in government activities at all levels of government. Lower levels of government with operational control over programs would then be able to select the most promising alternatives for implementation. However, obtaining a better governmental innovation strategy requires much more than greater decentralization. Governmental organization and elections need to be much more sharply focused so that bounded rational voters would demand better performance from their elected officials.

In order to obtain a more focused government, governmental functions and tasks would be reorganized into a four level government. The first two levels would remain the federal and state governments. Under the new design, the role of the federal government would be international affairs both military and political, basic constitutional human rights, the coordination of state activities, and the operation of programs with very large economies of scale. An example of an activity with very large economies of scale would be the research task of the separation strategy for governmental innovation. However, the federal government does not have economies of scale in operating programs for individuals and local concerns such as communities and small business. In such programs, most operational attempts of government to reduce individual risk through product safety, worker safety, and household safety should, therefore, be focused at the state level. The operation of most social programs, such as social security and welfare, would be transferred to the states.

Currently, the states vary enormously in size, population and resources. In order to make decentralization of governmental programs by the efficiency criterion uniform across states, the boundaries of the states would be reorganized such that the variations in size, population and resources of the various states would be greatly reduced. Under the proposed design, the role of the states would become promotion of the state economy and risk management where risk management includes welfare and the regulation of local hazards and the state environment.

Major reorganization would be required below the level of the state government. Currently below the the state governments are a large variety of local governments with overlapping jurisdictions. These local governments include municipalities, counties, towns and townships, and a variety of special purpose and hybrid arrangements. The fastest growing category of government is the special purpose district to provide services such as schools, fire protections, and water. Because voters are bounded rational and have limited resources to devote to ensuring public officials are held accountable, this complex structure should be simplified to two levels of elected government. These two levels of government in urban areas would be metropolitan and town government. In rural areas the government level corresponding to the metropolitan government would be the district government where the district would comprise a large area with common features such as the same agricultural crops. The government for local rural areas in districts would be the county.

The role of the metropolitan government would be to govern the local political economy with an emphasis on the production of metropolitan wide services. How much metropolitan governments would use special purpose governments to produce metropolitan services would vary greatly. The role of the town government would be to promote the community lifestyle. The roles of the district and county governments would correspond to their urban counterparts.

As technology and knowledge advance after the initial restructuring of government, the ideal governmental decentralization would gradually change. To maintain an ideal decentralization the professional review would be employed. In the operation of the professional review those activities which were constitutionally specified to reside at particular levels of government would stay at those levels, but all other activities would be controlled by the principle of decentralization.

Consider how the shift of those activities not constitutionally mandated at a particular level of government would occur. Governmental functions and tasks would be transferred from all subordinate governments to a higher government any time the higher government passed a legislative act for such a transfer. Such an upward transfer would not require a professional review, but could be challenged in a professional review suit by any of the subordinate governments involved. However, a decentralization professional review would have to be filled by a subordinate government wishing to decentralize a governmental function or task. In such a case depending on their analysis the judges could dismiss the case, decentralize the function or task, or grant the filing government an experimental variance.

In deciding whether a particular governmental task should be decentralized the judges would have to weigh the efficiency of decentralization against the value of additional variation through decentralization[5]. The economies of scale criterion translates to mean that government activities should be assigned to the level of government which can most effectively perform them. The knowledge criterion means that the less that is known about the consequences of alternatives for a government activity, the lower the level of government to which the activity should be assigned. The tradeoff between the economies of scale and the knowledge criteria requires that the value of future learning be compared with the value of economies of scale. There is, for example, some value in accepting a slight loss in economies of scale if placing the activity at a lower level generates much greater empirical consideration of alternatives and more imitation from a larger number of other governments at the lower level. The hidden cost of placing activities at too high a governmental level, especially without variation over time, is the loss of empirical knowledge about alternatives.

If the opinions were in agreement that a particular governmental task was being performed at the appropriate level the case would be ruled in favor of the defendant, the higher level of government. And if the opinion were in agreement that a particular governmental task should be decentralized, the task would be assigned to the appropriate level of government. For some activities the consideration of economies of scale will be sufficient to place an activity at the proper level, but for many activities, where the value of economies of scale must be weighed against the value of increased learning, reputable analyses could differ as to the most appropriate level of government. For governmental functions and tasks for which the some judges agreed that decentralization was appropriate, the subordinate governments would have the right to petition for an experimental variance.

In proposing an experimental variance, the subordinate government would have to demonstrate what could be learned by the variance. In some cases several governments at the same level might jointly propose an experimental design to obtain an experimental variance from the higher level of government. In cases where lower levels of government applied individually for variant roles, they would propose demonstration projects or pilot studies, and each of these individual pilot projects would have to offer a unique learning approach.

To encourage numerous experiments in government lower levels of government would have incentives to press for decentralization. If a lower-level government won a decentralization case, either by the decentralization criterion or by an experimental variance, the lower-level government would obtain the funds that would otherwise have been spent by the higher-level of government for performing the same task.

An example of a potential candidate for an experimental variances in the triple damages award for antitrust violations in civil actions. If a large firm, for instance, tries to bankrupt a small firm through predatory business practice, the small firm can sue the large firm in civil action to recover triple damages. The positive effect of this award is the creation of incentives for firms to avoid predatory practices. But at the same time, the negative aspect of this award is that triple damages may create incentives for small firms to file too many suits of questionable merit, that is, nuisance suits. With triple damages the national standard since 1914, we do not know the consequences that might result from double damages or from an award based on the inverse probability of being caught. Consequently, a state or groups of states could file for experimental variances in the amount of the award or the creation of a variable formula.

In the proposed design, a subordinate government could file for an experimental variance in the precedent of the judiciary one level above the subordinate government. Many court decisions involve difficult tradeoffs such as those between individual rights and the needs of society. When the court makes a single precedent, no empirical knowledge is gained concerning the consequences of possible alternative precedents. In an area of law undergoing rapid change, then, a multiple precedents would be useful to empirically determine the best precedent. Multiple precedents would be made operational through the professional review. If in setting a precedent the minority issues a minority opinion in conflict with the majority opinion, this minority opinion forms the basis for a subordinate government to file for an experimental variance in the majority precedent on the basis of the minority opinion.

The experimental variance applied to legal precedents would greatly change the dynamics of precedent setting when the majority of a court shifts from liberal to conservative or vice versa. Under the current operation of the judiciary the political shift of the majority of a court results in a corresponding shift in the political interpretation of the precedents. With the right of an experimental variance unless the judges were unanimous subordinate governments opposed to the majority would file for experimental variances. Thus a political shift in the majority would simply change which subordinate governments were filing for experimental variances. With more variation in precedents empirical knowledge of their effect on the political economy would be gained more rapidly.

Federal Role in Decentralization

The role of the federal government in the proposed governmental design would be to perform only those governmental functions and tasks within these functions for which the federal government were clearly more efficient than subordinate governments. Included among these functions would be most of the original functions of the federal government such as defense and foreign affairs. While the initial decentralization would transfer many operational tasks of the newer multilevel governmental functions to subordinate governments, the federal government would retain some operational tasks in most multilevel governmental functions. In addition, for all tasks performed by subordinate governments the federal government would perform the research and development task in the separation strategy for governmental innovation.

To illustrate why the federal government would retain some operational tasks in multilevel governmental functions consider environmental regulation. Currently it is more efficient for corporations to lobby for federal environmental regulations than to lobby for state environmental regulations. At the federal level they can reduce their lobbying costs and achieve a single national standard less restrictive than the standards adopted by the most environmentally activist states. Thus the current political criterion for environmental regulation create economic incentives for placing environmental regulation at the federal level.

The decentralization criteria would shift some but not all environmental regulation to the states. To take a specific example, acid rain is a national and international problem which cannot be effectively decentralized to the states; hence, the regulation of the environment would remain a joint federal and state activity. For example, it is hard to argue that the federal government has economies of scale in regulating the internal environment of houses and offices because the variation in state standards would provide much useful empirical information concerning the costs associated with the various safety standards. This empirical information would lead to better performance through imitation. Thus the federal government would retain control over such environmental problems as acid rain and other air and water pollutant flows between states and national borders. The states would have operational control over environmental standards and incentive systems within their boundaries as long as they meet the pollutant standards at their borders.

Similarly, while the regulation of most aspects of industrial safety would be transferred to the states, some aspects would not be transferred. Nuclear power regulation would not be decentralized because the consequences of a nuclear meltdown can be so catastrophic over such a large area.

However, the decentralization of governmental tasks would increase the role of the federal government in coordinating these tasks performed by subordinate governments. For example, as was pointed out decentralization of more environmental regulation to the states would increase the need for federal coordination. And in its role as a coordinator the federal government would be subject to a professional review as in all its other tasks. For example, under the professional review if the federal government tried in induce all states to raise the drinking age from 18 to 21 by threatening to withhold highway trust funds, many states would immediately file professional review suits for decentralization. At the very least some states would win experimental variances.

Also, for all decentralized governmental tasks the federal government would assume the research and development task for governmental innovation. Hence the federal government would continue to perform systematic experiments to test drugs and the impact of chemicals on the environment. Also, the federal government would perform more social experimentals to test alternatives theories underlying policy choices. But because of the risk factor not all policy alternatives could be systematically tested prior to implementation by subordinate governments. In these cases the federal government would systematically analyse the consequences of subordinate governments adopting different objectives and policy alternatives.

Whether or not decentralization would increase empirical learning about the decentralized government policies would depend on the choices made by subordinate governments. To be sure, decentralization would give subordinate governments a larger role in estimating the common weal; consequently, decentralization would increase the number of estimates of the common weal which could be selected simultaneously. In itself, however, decentralization would be insufficient to guarantee that the choices made by the states would exhibit enough variation to provide useful empirical information concerning alternatives, because every subordinate government could conceivably choose the same alternative. Variation would be likely to occur only if there were variations in political philosophies and a number of competing theories about how to achieve objectives.

Under the proposed design several factors would lead to greater variation in the estimates of the common weal and of the associated goals and means. First, as will be discussed the redistricting of states each decade would result in greater variation among the states. In addition, as freedom of location increases, individuals would be more likely to select a state, metropolis and town which would match their tastes, thus accentuating the differences among the subordinate governments. Because there would likely be more political variation among states than there is currently, the subordinate governments would be likely to vary in their estimates of the common weal. And assuming the current controversial nature of the social sciences continues for the foreseeable future, the selection of policies would also likely vary among subordinate governments.

While differences among states, in terms both of political philosophies and social theories, would produce considerable variation in policies, the amount of variation would likely be less than ideal in promoting empirical learning. Elected politicians would be interested in producing results by the next election, whereas the benefits of social experimentation would frequently be long-term. Thus incentives operating on elected politicians would encourage them to select the alternative with the greatest short-term performance. Alternatives which might produce much better long-term results, but which involve short term risks, would not likely be considered. In the proposed decentralization scheme, the responsibility for providing incentives for lower levels of government to empirically test alternative approaches would rest with the federal government. This is because under decentralization, the benefits in increased knowledge would be nationwide, whereas the consequences of accepting risk would be local. To encourage risk taking by lower levels of government, the federal government would provide subsidies to induce lower levels of government to try new alternatives. The federal government might fund part or all of the proposed experimentation, or might even provide a bonus to states for trying a new approach to solving a particular problem. To reduce its costs, the federal government would encourage lower levels of government to apply for experimental variances. For example, it might be more cost effective to subsidize a major metropolitan government in applying for an experimental variance to test a new alternative than to subsidize a state government for the same thing.

State Government: Organization

As the political economy is currently constituted, the wide disparity among states in terms of size, population, and resources implies that for any decentralization scheme based on economies of scale, decentralization from federal government to the state governments would vary from state to state. A simpler decentralization system would require that the states be reorganized in such a way as to promote a uniform decentralization.

To determine the appropriate size of the reorganized states, the economies of scale in business as well as government must be considered, since complying with the greater variation in state laws resulting from decentralization would mean higher costs for producers due to some loss of economies of scale, although these costs would in time decline with the growing flexibility made possible by automation. To avoid this greater variation altogether, however, would be to incur the cost of foregone empirical learning about how to continually create a better political economy.

Achieving a compromise between these costs means that states must be made large enough to comprise a market with economies of scale and at the same time must have economies of scale in operations of government. In order to meet these requirements, each state should be large enough have at least one research university and a sufficiently large professional talent pool that would enable the state to form its own opinions about risk management. Bearing these stipulations, this book proceeds on the assumption that the various states should contain a population between 5 and 20 million people. Given our current arrangement of states, then, several of the less populated states would have to be combined into larger units.

The method proposed here for maintaining the states with sufficient resources and size in an environment of constantly changing patterns of technology and state growth is to redraw state boundaries each decade[6] Since congressmen would be interested primarily in redistricting that would maximize their reelection chances, the nationally elected representatives, the senators would draw up three alternative plans to be submitted to the voters for choice. These alternatives would have to be approved by the President and could be subjected to a professional review before being submitted to the voters. The criterion for redistricting should be cluster analysis, a statistical technique used to group entities with similar attributes, to create states with internally homogeneous interests. Thus organized, states would promote homogeneous interests in such a way as to satisfy the general benefits criterion of the professional review.

A consequence of reorganizing the states every decade would be a greater difference in the political economy of the states than currently. Reorganization each decade to create states with internally homogeneous economic interests would imply greater heterogeneity among states. Also, with increasing freedom of location of individuals, differences in political philosophy would also be accentuated. For example, the states have the right to file for experimental variances on the rulings of the Supreme Court. Consequently, states with a liberal majority would file for the liberal minority precedent when the court majority was conservative and conservative states would take similar action when the majority of the Supreme Court was liberal. Thus with freedom of location states would tend to attract individuals and households with similar views.

These heterogeneous political economies would evolve different arrangements for state government. This is because given the principle of decentralization, the task of organizing state government would fall to the individual states. Some would maintain the more traditional two-house legislature while others might modify the legislature to parallel the proposed federal legislature. Similarly, the manner in which states implemented the concept of a professional review of government would vary considerably. As the state governments would be assuming many operational considerations from the federal government, they would be delegating many government responsibilities to the next level of government, the metropolitan governments.

Nevertheless, a common theme among the state governments would be the development of more technical capability in order to perform in their expanded roles. This means that all states would move to full-time legislatures with professional salaries such as California and New York currently have, and all states would have to upgrade the quality of their professional staffs in the state bureaucracies. Such a move would be a projection of current trends[7].

State Government: Role

Under the new design, the role of the state government is to promote the state economy and to take operational control over most government programs for individuals, such as social programs and risk management. Because revenues would be redistributed according to this redistribution of legislative responsibilities, the states would have the resources to fund the activities acquired from federal control. This redistribution of funds and responsibilities, however, places more of the burden of estimating the common weal on the shoulders of the state governments. Thus the state governments would face more difficult problems than currently.

One such problem would be the promotion of the state economy. Traditionally states have promoted business through investments in the infrastructure, for example, by financing roads, schools and amenities, such as state parks which also promote economic activity by attracting a quality workforce. States would continue to perform many tasks in these functions and through the decentralization criterion delegate the remainder to the metropolitan governments.

One current method of promoting economic activity would be sharply curtailed. Today, states and municipalities try to attract business by offering tax and other incentives and as should be expected, businesses use this state and local competition to their advantage to obtain the best deal. Empirically these incentive plans have proven ineffective in promoting economic activity[8]. Given the professional review, such practices would be successfully challenged both on the grounds of efficiency and lack of general benefits.

The aspect of economic promotion which would present the states with an extremely difficult problem would be efforts to accelerate the transfer of technology from research to the marketplace. The problem as was pointed out in Chapter 1 is that basic research is a public task funded by government and invention is a privately funded task. The difficult problem is knowing where to draw the line between public and private promoting of research in the continuum between basic research and new products. To promote better government innovation the federal government would fund only basic research and general applied research leaving the state governments to fund all activities directed towards transferring federally funding research to the private sector. A state which develops a better means of transferring knowledge from the public research to private startups, would enjoy increased prosperity by promoting new types of industries displacing older industries. A state which completely foregoes such promotion would forego the benefits of expanding industries. At the other extreme a state which funds too many promotion attempts would likely end up with expensive boondogles.

A second difficult problem each state would face would be welfare and social programs. In the initial of governmental functions and tasks most social programs such as social security would be transferred to the states. Thus each state would be responsible for retirement programs, medical programs, unemployment insurance, and programs to aid the poor. Consequently, each state would have to make its own estimate of the appropriate level of social and welfare programs. Consider medicine. Currently the federal government has medicare for the elderly, medicaid for the poor. Individuals with fringe benefits have medical coverage through their employers, but millions of individuals do not have medical coverage and do not qualify for either medicare or medicaid. Local governments provide some medical care for individuals lacking medical insurance. In the proposed decentralization the states would assume complete operational control over medicare and medicaid. Thus each state would have to make the difficult estimate of what level of medical care each individual would be entitled. In making this estimate each state would have to select somewhere between the extremes of only the medical care each individual could privately afford to the best the medical profession is capable of providing. Currently one trend in social programs is for the states to make them required programs of private firms for employees. As informational society develops such a trend would be reversed. In informational society most individuals would find employment through a sequence of short term contracts. For this reason it would be efficient to shift various fringe benefit programs such as medical insurance from the firm to the individual in order that each individual would have consistent long term coverage.

Consequently, states would have incentives to use social inheritance as a method of financing social programs. As was pointed out in Chapter 4, each year every citizen receives his share of capital from those who died that year. The federal government would collect from the estates of the deceased and pass these funds to the states for social inheritance implementation. In creating the state social inheritance programs a state could make certain types of insurance mandatory. Consider medicine, for example. As indigent care is funded at public expense, the public has an interest in making certain that all persons carry some form of medical insurance. Such a requirement that all residents carry a prescribed level of medical insurance funded from the income of social inheritance, is simply an extension of the current requirement that all automobile owners carry auto insurance.

Faced with the assumption of the retirement income provision of social security states might place heavy reliance on social inheritance. Prudently managed, social inheritance could provide a guaranteed income which would increase over the life of the individual. In implementing their new welfare policies, state governments would have to decide how an individual could invest his social inheritance. As individuals have very different preferences concerning the tradeoff between immediate and future income, different individuals would want to make very different types of investments. However, allowing individuals unlimited freedom to consume or invest their social inheritance, especially if individuals could sell the discounted value of future social inheritance, might result in a large number of destitute fools in old age requiring public support. Thus, each state would face the difficult decision of deciding what are the appropriate restrictions on the use and investment of social inheritance.

A third difficult problem for each state would be the operational control of risk management for individuals and local concerns. Governmental tasks which would fall into this category would include the governance of worker safety, local environment, drug release policy and the previously discussed social inheritance. Experience with federal risk management since the 1960s has driven home the point that there is tradeoff between reducing risk and economic activity especially economic innovation. Each state would have to make the difficult estimate of the common weal in formulating risk management policies.

States in seeking to balance risk reduction against other social concerns would be subject to the professional review. As was pointed out, new risks are subject to screening while old risks are subject to standards. Politically it is much easier to prevent new risks than to control old risks, hence risk management is subject to a double standard which inhibits the development of new technology. However, the current practice of discriminating between old and new risks would probably not stand a professional review challenge of consistency. All states would have to seek better approaches to equate the dangers of old versus new risks. Some states might try proposed approaches such as a methodology of comparative risk. With decentralization, more than one approach to risk management is likely to be considered empirically at one time. Under the impact of the professional review states would seek a gradual reduction in the asymmetric treatment of new and old risks.

In trying to balance risk management between alternative risks and to adjust to changing technology each state would face the difficult problem of selecting the best combination of regulatory approaches for each risk. The original method of regulating risks was through common law. This method will be called the old incentive approach because common law precedents created incentives for creators of risk to take appropriate action. Included under the term old incentives will be the incentives to avoid civil and criminal suits under statutes imposed over common law. The middle approach starting with the Progressives was the bureaucratic approach wherein government agencies created detailed regulations which were monitored by government inspectors. The newest approach which we will call the new incentive approach is to create economic incentives for the participants to reduce the risk. An example here is the incentives for large firms to improve worker safety in order to reduce their insurance premiums.

In informational society given the forecast for an even more rapid rate of technological advance, innovations in risk management would likely place a much greater emphasis on creating incentives to promote the participants to reduce the risks themselves over time. One reason is because properly designed incentives would promote a quicker adjustment to a rapid rate of technological change than a state bureaucratic approach.

Also the rapid advance of micromonitors and the social nervous system would encourage innovations to reduce the costs associated with risk managements. For example, an innovation in all types of risk management would be the creation of decision support systems for each type of risk management. Under operational information policy one goal of this decision support system would be to ensure that all participants had complete current knowledge of the risks they faced. For example, with the development of inexpensive micromonitors the environment will be constantly monitored and the readings would be available to all parties. A second aspect of the creation of decision support systems would be to reduce the costs of determining the status of all laws and administrative rulings. User friendly software would enable business decision makers to quickly determine exactly what regulations applied in certain sets of circumstances. Also, current delays caused by bureaucratic discretion would be reduced by the creation of expert systems.

The fourth difficulty the states would face is the difficult problem of balancing concerns for economic promotion, social programs and risk management. Consider first medicine. Medical research has created effective medical procedures which society can not afford if unlimited medical care becomes every individuals right. Consider next the difficult task of risk management for which the states would develop policies balancing the need to reduce risks, such as safety hazards, environmental damage and loss of social inheritance against other considerations such as cost and a higher rate of economic activity, private invention and innovation under unregulated markets. In this balancing we shall assume that in informational society reducing risk almost always entails social losses such as higher costs or inhibiting private enterprise. Thus risk management requires the states to make an estimate of the common weal in the form of the desirable balance between opposing social goals.

What makes the balancing between economic promotion, social programs and risk management extremely difficult is that decentralization of programs from the federal government will mean that governmental tasks will be performed in a more competitive environment of the states as opposed to nations. Today, state competition exists in an environment of free trade, where no state can create money, and where state governments are generally required by state constitutions to balance their budgets. The competition among states is thus more intense than among nations because the states have fewer devices with which to impede competition. In informational society because individuals and firms would have greater freedom of location, competition among states to promote their economies will intensify even more[9].

In making such estimates the increased competition among states does not mean that the states would suddenly curtail all social programs or abandon risk management to the unregulated market. With greater competition states would have greater incentives to fund social programs as part of the social inheritance program especially if the level of social inheritance increases with software becoming an increasing component of national income. Thus states which used social inheritance to fund social programs could maintain a positive business climate.

Also, increased competition would not mean an abandonment of risk management. Consider, for example, environmental regulation. With advances in chemistry and biology man will continue to introduce ever new substances and lifeforms into the biosphere. With the rise of biotechnology, such as the production of compounds using genetically altered bacteria, the number of trained professionals aware of the interaction of human activities and the biosphere will increase and some will inform the public of the environmental dangers resulting from such technology. Operational information policy will make such information available to residents gradually learn the lesson that preventing an environmental disaster such as contaminated ground water is very much cheaper than cleaning up such a catastrophe afterwards. As concerns such as those surrounding the extinction of species and the loss of the ozone layer are likely to increase, environmental regulation would increase over time in all states, but with considerable variation between states. Greater competition in environmental regulation would result in innovations to reduce the administrative costs of the participants.

State Innovation Strategy

State governments would need considerable empirical knowledge to innovate successfully in their difficult tasks. For each of these tasks competing factions would propose conflicting objectives as estimates of the common weal. And experts would have numerous conflicting policy proposals to achieve each of the various objectives. Hence, the ability of the state governments to innovate depends on their ability to accurately predict the consequences of selecting a particular objective and associated policy. Because the social sciences are likely to be characterized by numerous controversies for the foreseeable future, accurate knowledge of the consequences of selecting an objective or associated policy would require an implementation of that selection.

States would obtain some empirical knowledge of their alternatives from two sources. First, variations in the objectives and policies implemented by the various states would provide useful empirical knowledge. Second, the federal government in its research and development role would perform experiments and pilot studies. In this section we shall discuss how the proposed design promotes state government innovation by increasing empirical learning.

An important consequence of the proposed decentralization would be more variation in the estimates of the common weal. Consider, for example, managing the risks of drugs. Granted, it is efficient for the federal government to test new drugs and determine their side effects; however, having the federal government set a single nationwide standard for risk greatly limits empirical knowledge of the correct balance of the two types of risk in drug release into the marketplace. The first of these is the risk from some unforeseen, dangerous side effect. The greater the testing of a drug the further this risk can be reduced. The second risk is that the longer the testing period, the more individuals would be denied the positive effects of an effective drug. Currently individuals journey to foreign countries in order to obtain the benefits of such unreleased drugs, and critics claim that the federal Pure Food and Drug Administration errs by placing too much emphasis on reducing the first type of risk while paying too little heed to the second type. The estimate of what balance between these two risks is in the common weal is a judgmental, not an analytical decision. Therefore the risk management policy of the appropriate amount of testing would be decentralized to the state in order to obtain a better long run estimate.

In the proposed drug testing and release program, the federal Pure Food and Drug Administration would test the drugs and create and maintain databases with complete descriptions of the side effects of drugs and status of their testing. These databases would then be accessible to the individual states and all interested parties. The more rapidly states released promising drugs, the greater the state would need a mechanism to warn potential users of side effects as they are discovered. States would implement their drug release policy by using the data support system maintained by the federal Pure Food and Drug Administration. Most states would probably establish a drug administration to establish a release policy based on the amount of testing completed by the federal government. This release policy would be integrated with the decision support system used by doctors to prescribe drugs. Expert system advances in medicine will eventually create programs which will select the best drug for a patient's illness conditional on his past medical history. A state could integrate the drug release policy into these systems so that the physician would be informed, for instance, that a drug was available without restrictions, required informed consent to the specified dangers, or required consent of the drug release agency.

In formulating its drug release policy, each state would have to estimate the socially desirable balance between the two types of risks. Given the forecasted differences among the reorganized states, state estimates of a desirable balance between the two types of risk would vary considerably. Part of the federal government's role of research and development in innovation would be to have the empirical consequences of these variations systematically analyzed. Over time through this empirical knowledge a better estimate of the common weal in the balance between the two two risks in drug release would emerge.

Another governmental task for which variations in the estimates of the common weal would be very beneficial is the government's role in providing medical services. Medical research has created medical procedures which are effective but which are so expensive that providing all individuals unlimited access would be extraordinarily expensive. Variations in both the level of care to which individuals are entitled and the methods of providing medical services would be essential for innovations in the production of medical services.

A second important consequence in the proposed decentralization is that alternative policies to achieve social goals would be implemented simultaneously rather than sequentially. Consider the problem of selecting effective policies to manage the environment. Under the proposed decentralization of environmental management, the federal government would be responsible for environmental research and environmental coordination among states by specifying air and water quality at state and federal boundaries. With this coordinating power, the federal government would control such problems as acid rain and pollution of coastal waters. States would thus have operational control over how these federal boundary standards were to be achieved and, moreover, would assume full control over local pollution.

Giving states full control over local pollution increases the likelihood that alternatives policies would likely be tested simultaneously rather than sequentially. To illustrate why decentralization would lead to a much more rapid testing of alternatives consider the evolution of federal environmental policy from 1970 to today. There are a wide variety of administrative and incentive approaches to controlling pollution[10]. One administrative approach is technological-based emission standards and examples of incentive approaches are emission taxes and market incentives. One proposed market approach is the concept of pollution rights. For example, given an environmental target such as so many tons of a particular pollutant released per year in a particular metropolitan area, firms would bid for the right to emit a percent of the target pollution level for a locality. The purchaser could subsequently sell this right.

The federal clean air and water legislation of 1970 and 1972 mandated the states achieve the federally designated standards using only one approach, technological-based emission standards. The states were given little discretion in setting the standards or in devising alternative methods of achieving the standards. In 1977 amendments were made to make this legislation more effective[11]. Now in the 1990s under the Bush administration environmental risk management is finally moving to test the effectiveness of implementing pollution rights, a market incentive approach.

The centralized approach to environmental control illustrates that as long as two decades can ensue before the alternative approaches are empirically tested. However, under decentralization of environmental risk management these alternatives would probably have been tested simultaneously. Environmentalists generally dislike the market approach out of fear that once pollution rights are established they will be hard to adjust. Thus liberal, environmentally oriented states would probably have opted for technological-based emission standards. In contrast, conservative, market oriented states would probably have opted initially for market incentive approaches to pollution control.

Nevertheless, decentralization in itself is insufficient for there to be variations in the policies among the states. For example, if the political majority of each state had the same views there might be no variation whatsoever. However, the reorganization of the states each decade and greater freedom of location made possible by technological advances would accentuate the differences in the political economies of the states. Thus in an political environment where political factions dispute ends and experts dispute means decentralization of tasks to the states would result in considerable variation in both ends and means. The empirical consequences of such variations provide the empirical knowledge for innovation as states would imitate successes and abandon failures.

The systematic analysis of these variations in state goals and policies would be part of the federal government's research and development task in state government innovation. Scientific information policy and federal research grants would provide the observations and financial support for this systematic analysis. The federal government would also fund experiments and pilot studies advancing state innovation. Because of cost considerations most of these experiments would be conducted on much smaller institutions than the states.

Consider, for example, medicine which would become one of the more difficult problems for the states. Most federal research in medicine, up to now, has been directed developing effective medical procedures without any consideration of the cost. The public has come to expect that these procedures should be generally available to all patients. In informational society the federal government would devote considerably more research effort at alternative efficient methods of medicine. For example, the federal government would fund research to investigate the positive effects of various type of preventative medicine. Given the cost considerations most experiments would likely be performed on towns not the states.

In the area of risk management the thrust of the federal government's research and development efforts would be to create incentives for the participants to reduce the risks themselves. These incentive systems would be based on operational information policy which in itself would create incentives by making all the participants in a risky process better informed of their respective risks. For example, consider decentralized environmental risk management. As technology advances the cost of monitoring the environment in great detail will fall. The federal government part of its task in providing voters with detailed comparisons in subordinate government performance would provide residents detailed profiles of pollution in real time in their homes, workplaces and local environments. Subordinate governments and private parties would make additional measurements. Already, the concept of an environmental right-to-know is moving information policy in this direction.

Consequently, because voters would be much more aware of environmental hazards, voter pressure to enforce environmental laws would be continual instead of varying with environmental catastrophes as is currently the case. In addition, private parties and local governments would pursue a much greater number of civil suits to enforce pollution standards. Faced with greater public awareness and a greater probability of a civil suit, many firms would seek the positive public relations of maintaining a good pollution record. Finally, the states would have much stronger incentives to constantly innovate in environmental risk management.

However, some types of risk management would require extensive research in order to create better incentives. Consider worker's safety[12]. The original worker safety risk management was common law. In the progressive period states created workman's compensation programs which provide workers immediate compensation for injuries without having to pursue a civil suit. In the 1960s the federal government added the Occupational Safety and Health Administration, OSHA, to establish and enforce worker safety standards. Under the proposed decentralization the federal role would be reduced to research and development of safety programs and OSHA's role of establishing and enforcing safety standards would be delegated to the states.

To forecast how the federal government research and development task would promote greater innovations in safety, consider the limitations of the current worker safety system. The current worker safety system creates few incentives for the participants themselves to promote safety. The small number of firms for which safety is regulated by common law have economic incentives to avoid negligence in worker safety. Large firms have economic incentives to promote safety in their firms to qualify for reductions in their workman's compensation insurance premiums[13].

Indeed, under the current worker safety system most of the incentives are perverse. Since workers do not pay for workman's compensation insurance they have no incentives to promote safety in order to reduce the insurance rates. And in hard times workers have incentives to file dubious claims[14]. Firms have incentives to lobby state governments to the reduce the workman's compensation benefits to workers and to lobby the federal government to reduce the number of OSHA inspections and the magnitude of the fines for safety violations.

In informational society states would have the difficult task of deciding whether to promote safety through establishing and enforcing worker safety standards or trying to create better incentives such as by empowering employee committees with the right to act to monitor and promote safety at their workplaces. As advances in the technology would create opportunities to design much better incentives systems, many states would opt for this approach. And the federal government in its research and development role would fund pilot projects.

For example, consider how advances in the social nervous system would enable the insurance-rate-reduction incentives to be extended from large firms to small firms. The problem is the size to the sample required to demonstrate a better safety record. Considered individually small firms are too small to establish what variations in their safety program are statistical significance in improving worker safety; therefore insurance firms have no valid procedure for granting insurance rate reductions to small firms who have superior safety programs.

The type of pilot project to establish a valid statistical procedure for granting insurance rate premium reductions to small firms would be to have the trade association in conjunction with the federal research and development agency, state insurance board and insurance industry representatives to systematically test safety alternatives across the firms in the industry. Consider just one example worthy of testing. Currently rather than test workers for drugs some firms are testing workers each day for eye hand coordination. The pilot project would systematically test the use of such a machine across the industry. If the use of such a machine were statistically demonstrated to reduce accidents then its use would qualify firms for a rate reduction. The pilot project would also experiment with various combinations of splitting insurance premiums between firms and workers. Thus through systematic safety experimentation throughout an industry insurance rates could be based on installed safety equipment, safety training and programs.

What currently inhibits such an approach is the administrative cost of obtaining, maintaining and analyzing detailed records on each firm. As the social nervous system advances in information society insurance firms the administrative costs associated with such detailed records would fall greatly. Also given operational information policy insurance firms would have access to all safety features such as safety equipment, safety training and programs installed in each of their client. With variations among the states operational informational policy would also give insurance firms the right of private inspection. Such a right would privatize safety inspections.

The federal government in this research and development role would be much more effective at promoting safety than the current OSHA. For example, to reduce nerve damage from continuous, high speed input at computer keyboards OSHA is likely to be more effective promoting research and development to shift to voice recognition input than trying to define standards for existing manual keyboard input. Once voice recognition input becomes more efficient than typing input taking into consideration reduced insurance premiums, market incentives will shift technologies eliminating the old nerve damage problem. Also, transferring operational control to the states will mean greater policy variation leading to a faster rate of discovery. Variation is needed to discover the value of creating better incentives for workers to promote safety themselves as well as better incentives for employers.

Metropolitan Government

Currently the political economies of most urban areas are a complex pattern of overlapping jurisdictions between various types of local governments such as county, town and township, municipal, school district, special district such as a municipal utility district, and private quasi-governments such as residential community associations. In these political economies, the political units arrange for the production of public goods by a wide variety of public, private and joint arrangements[15].

In informational society, the provision, which is defined as the arrangement for the production, of local public goods would be consolidated into two levels of government[16] for two reasons. First, as the voter has limited cognitive skills and resources to hold publicly elected officials accountable, a reduction in the number of governments would improve political performance. Second, as will be discussed economic incentives would be superior to political incentives for obtaining better local services. Nevertheless, the arrangements for the production of public goods would remain complex.

In urban areas[17], the metropolitan government would be responsible for arranging area-wide services such as water, electricity, and sewage. Under the jurisdiction of this metropolitan government, would be the towns which would be responsible for arranging local services. The metropolitan government would correspond to the standard metropolitan statistical area, SMSA, which, given regional redistricting, would generally lie within a single state. Creating this metropolitan government in most current urban areas would require consolidation of the area-wide responsibilities into the single metropolitan government. Once created, the criterion for the jurisdiction of the metropolitan government would be dynamic, like the present SMSA criteria, in order to cover the land area that was primarily integrated into the economic life of the city. This territory would be adjusted annually. Towns would be governments for much smaller groups of from several hundred to several ten thousand individuals with compatible lifestyles within the metropolitan jurisdiction. Creating this local government arrangement in most current urban areas would require partitioning the larger cities into smaller towns.

To further elaborate the division of functions between the metropolitan and town governments, let us review the previous discussion on the difference between their criteria for land use. At the metropolitan level, land use would be regulated by market considerations as modified by zoning. The metropolitan government could zone its various metropolitan regions into areas for specialized activities, such as the central business district, metro parks, and industrial parks but it would have to be able to demonstrate the rationale for this zoning. At the town level, however, the government would function to support the lifestyle of the community. Thus, the town would have the right to regulate land use outside the home according to the social criteria of the community, which in some cases would be market criteria and in some cases not. To promote its chosen lifestyle, for instance, the town could prohibit bars, massage parlors, or any other facilities for activities the town desired to prohibit. If a town zoning ordinance were challenged, the burden of demonstrating that the ordinance denied a basic right would rest with the challenger.

To make feasible such a division between zoning powers of the two levels of government, the metropolitan government would have to directly control land outside the jurisdiction of any town. Metropolitan land would thus be the mechanism for ensuring economic freedom in the traditional sense. If an activity were legal, the metropolitan government would have to allow an individual the right to pursue the activity on land purchased in the direct metropolitan jurisdiction. This right would extend even to controversial businesses such as massage parlors and porno shops. In the towns, however, the individual would not necessarily have this right.

A consequence of this division in land use would be that while most towns would have a small business district for local businesses and offices of people working through the social nervous system, almost all major business activities would be located on land directly controlled by either a metropolitan or a district government. On this land standard commercial practice would apply making the conditions for conducting business reasonably predictable. Businesses located in a town, however, would be subject to the zoning practice of the town, which could shift in accordance with a shift in local politics, thus making the conditions for conducting business less predictable. For example, a new town government could greatly change the conditions of access by outsiders to the town. For a business not focused on the local market, then, locating in a town would add a risk not found in locating on metropolitan land. Inasmuch as the professional review would prevent towns from offering new businesses services below cost, there would be no economic incentive for other than local businesses to locate on town land, since locating on town land would not free a business from metropolitan taxes.

Under the proposed division between town and the metropolitan governments, the metropolitan governments would be large enough to shoulder the responsibility for many governmental functions from the states. Indeed in the proposed design not only would many governmental tasks be decentralized from the federal government to the states, many current state governmental tasks would be decentralized to the metropolitan governments. For example, metropolitan governments would assume greater educational responsibilities such as providing four-year-teaching colleges and adult retraining for new jobs. And some states would decentralize medical care to the metropolitan governments.

However, many government activities such as education would continue to have components at all levels of government and a few activities would be moved up to the state level. Since each state would be responsible for determining risk policy towards the individual, states would control such matters as building and fire codes. This move would reflect the tenet that as technology transforms building techniques and materials, a larger pool of professionals is required to formulate policy governing building safety and a larger market area with uniform standards is needed to promote building automation.

Also, in keeping with the principle of decentralization, the organization of metropolitan government to perform its assigned governmental functions would be determined by a metropolitan constitution subject to the approval of the voters but not the respective state government. Because the metropolitan governments would be larger than current city governments and would be subject to a professional review, it is assumed that the best type of metropolitan government would be a checks-and-balances type government based on separation of powers with a strong mayor as executive, a council as a legislature, and a metropolitan judiciary. These government officials would be full-time officials with salaries compensatory with other opportunities in the political economy. In keeping with the bounded rational principle of a short ballot, judges and officers in the public service corporations would be appointed by the mayor and confirmed by council. Numerous variations in metropolitan government would exist. For example, one factor would be that the number of special districts which currently exist in urban areas would be retained to provide public services. Also, some metropolitan areas would maintain elected judges and elected special districts. In addition, the power of the mayor would vary widely among metropolitan governments.

Regardless of how the metropolitan governments were organized, they would be very concerned about the efficiency of their services. As individuals and firms experienced increasing freedom of location, metropolitan governments would have to compete more and more vigorously to attract new industry and to hold existing plants. To attract new industry the metropolitan government would have to simultaneously provide low-cost services and promote a quality lifestyle. And with the delegation of most lifestyle issues to the towns, metropolitan governments would focus on the efficiency of their services.

The search for efficiency at the metropolitan level will be primarily focused at how to create better services at lower cost and how to create a more efficient regulatory apparatus. Consider first the problem of regulation. Metropolitan regulation would cover metropolitan services such as restaurants. Like the states, the metropolitan governments would have a wide range of alternative arrangements for regulation involving various combinations of administrative control, incentives, and the search for efficient property rights for regulation. Like the states innovation in metropolitan government regulation would frequently involve using information policy to create better incentives. For example, the metropolitan governments could simple release inspection reports of the restaurants to the consumer services or provide the consumer services an information right for inspection. In the later case the consumer choice aided by consumer service analysis would regulate restaurant cleanliness.

Consider now the metropolitan governments' search for efficiency in the provision of public services. Some readers may have the misconception that there is a clear division between the public and private provision, production and consumption of goods. In actuality there are a large number combinations with varying degrees of public and private participation. For example, governments can arrange to have public goods produced by private production through a variety of arrangements such as contract, franchise, grant or voucher. Governments can also arrange to have public goods produced by other governments.

The best alternative method of production is a function of the attributes of the public good. Efficient private production of public goods requires that the good be clearly defined, that there exist numerous competing firms, and that government is effectively able to monitor the mechanism selected for private production. If the private firm producing a public good is to be paid directly by the customer, the private firm must have the ability to exclude nonpaying individuals from consumption. Selecting the best alternative is far from simple given the large number of alternatives and the possibility of mixed forms of production.

In addition, efficient provision of metropolitan services would necessitate that metropolitan governments rapidly adapt to advancing technology. Such advances would create opportunities for decentralization. For example, one metropolitan service likely to be either decentralized to the towns or even privatized completely is the library service. As laser disk capacity increases and costs decline, most households could maintain a small library, and an individual could obtain any reference material from private information utilities. Moreover, advancing technology such as the emerging social nervous system would provide metropolitan governments with numerous opportunities for innovation and create new problems. The social nervous system could be employed for monitoring utility meters, monitoring the environment, and coordinating police protection with private surveillance of private property, but the social nervous system would also create new opportunities for criminals to exploit this medium.

Technology would also enable metropolitan governments to experiment with new solutions to old problems. Consider, for instance, the provision of electricity. Because of the problem of natural monopoly cities generally provide electricity either through private utility companies under regulation or through city owned services. One approach to improve city owned utility services would be to experiment with management situations markets to achieve efficient service. As management becomes increasingly analytical and visual communication becomes inexpensive, the management of a facility can be achieved remotely with occasional onsite inspections. Management groups over a wide area could compete for any opening in the management situations market. The criterion for the competing management groups would be output-price minimization consistent with a fair rate of return on equity[18]. An alternative approach to providing electric power is greater reliance on competition between power producing firms. A compete market approach would be to create a market for electric services by treating the electric grid as a common carrier[19]. When, or perhaps if, superconductivity becomes economical, the number of potential suppliers in the electricity market would become very large.

Currently the rage among conservatives to improve governmental efficiency is privatization of governmental services. Creating a framework such that privatization of governmental services which form natural monopolies would promote efficiency is not always possible. In informational society the social nervous system and freedom of location would make the use of the management situations market an attractive alternative to regulate natural monopolies. Such use would lead to much greater efficiency than the creation of large numbers of politically elected special districts. Thus for reasons of both efficiency and accountability these special districts would be folded into the metropolitan governments of informational society.

Given the numerous opportunities for metropolitan innovation, a key factor in the rate of innovation would be the strategy for implementation. Because the number of metropolitan governments is likely to remain large and these governments would have similar problems, adoption of the separation strategy would improve innovation performance in most cases. According to this strategy the federal government is responsible for research that promotes discoveries leading to innovations and for funding pilot studies and other empirical tests of new alternatives. To provide a focal point for applied research and to train more competent professionals research in metropolitan services should be organized in selected professional schools at research universities. With increasing freedom of location for individuals and firms, the political philosophies and theories of the metropolitan regions would diverge even more than those of the states. This, coupled with federal subsidies for experimental trials of new alternatives, would lead to a very active environment for innovation. A fundamental split between conservative and liberal approaches to metropolitan problems would occur over how much the innovations would take the form of new ways to privatize public services.

Town Government

The function of the lowest level of government–the town–is to promote the lifestyle of the majority in order to develop a greater sense of community in each town. In informational society the town is intended to promote personal relationships to counterbalance the impersonal, competitive institutions outside the town. To the extent that the towns succeeded in creating communities, the metropolitan-town organization would promote economic efficiency, because the residents would participate in many local activities, which would reduce the social costs of transportation.

As was pointed out in Chapter 5, the two powers the town would use to promote the lifestyle of the majority are control over the physical organization of the town and control over activities outside the residences. Under these conditions, households choosing residences would tend to simultaneously create a wide variation in the lifestyles among the various towns and compatible lifestyles within each town, to the extent that households had freedom of location. For example, towns of information workers might well restrict the flow of automobile transportation to promote human interaction, whereas a town in which the majority needed to travel to work would likely maintain a rapid access automobile transportation system. Also, a town of fundamentalist Christians promoting family life would permit very different activities than a town of gays.

The political organization of towns would also vary considerably. Some small towns would be organized as direct democracies with all adults voting on all the issues in the town hall. Larger towns would be more likely to use a form of representative government such as a council and town manager. Since the judicial system is based on judges in law and judges in fact, the economies of scale argument would place the lowest level of courts at the metropolitan government. The great variation in size, political organization, and lifestyles of the large numbers of towns would generate a tremendous variation in the demands for local services.

The metropolitan-town dichotomy would promote the current trend for local governments to use a wide variety of alternative arrangements for supplying services. To provide local services a town could decentralize the service to private associations or individuals, could produce the service itself, or could buy the service from another government or private firm. In addition, many services could be provided by joint production arrangements between different levels of government or public and private arrangements.

The clause of general benefits, stipulating that government activities should be self-financing if self-financing is efficient, would create incentives for towns to decentralize to associations those activities which are easily self-financed. Although the town would have the power to promote belief systems, including religious beliefs, towns would very seldom use local tax money to fund churches and other institutions promoting belief, because churches have been traditionally financed by the parishioners and the self-financing clause of general benefits would maintain that relationship. In addition, sports and arts and crafts activities would frequently be financed by associations.

Currently city and county governments contract a wide variety of services from private firms[20]. The list includes services associated with many types of public facilities such as the construction, maintenance and sweeping of streets; the construction, operation, and custodial services of community centers; and the construction, operation, and maintenance of recreation facilities such as golf courses. Local governments also contract various types of inspections such as building, mechanical, electrical, health and plumbing. Moreover, private firms sell various types of administrative services to towns such as data entry and processing, tax assessing, and tax and utility billing and processing. In addition, towns purchase professional services such as architectural, auditing, legal, and management consulting. Finally, local governments contract private firms to produce many services such as ambulance, animal control, crime prevention and patrol, fire prevention and suppression, health, landscaping, parking services, and social services.

The development of the metropolitan-town organization and the social nervous system would create incentives for the development of an increasing number of firms producing services for towns. The creation of much smaller towns would create incentives for entrepreneurs to create additional professional services for towns to solve specific technical problems such as recommending the best combination of technical surveillance and human police protection. Moreover, the great increase in the number of towns increases business opportunities by providing a large increase in the number of potential contracts. Finally, the development of the social nervous system and the shift of services to the social nervous system greatly extends the area over which a firm can compete.

In the decision regarding whether to produce or buy the service increasing numbers of towns would opt to use the market mechanism. The efficiency of the market mechanism depends on the number of demanders and suppliers as well as the information policy affecting the participants. With the metropolitan area partitioned into small towns, there would be a large number of demanders and as services shift to the social nervous system, the number of potential suppliers would greatly increase. Thus, many markets for town services with large numbers of suppliers and demanders should approach the conditions of economic efficiency. Under such conditions, some towns would become public counterparts of the hollow corporation, that is the town would have a very small staff arranging all public services as contracts.

To effectively exploit advancing technology, the political economy of the metropolitan area must constantly innovate. For the state and metropolitan government a recommendation was that these governments switch to a separation strategy for innovation. And given the very large number of local town governments, this strategy would be even more effective for the towns. As was pointed out, the federal government assumes responsibility for research and development, and operational aspects of public services are assigned to the level of government which can provide the service most efficiently. As a consequence, operation of local services would frequently be split between the town and metropolitan governments.

The market for town services would promote active innovation. First, there would be a lot of variation in the production of town services. Some towns might have private fire departments and others public, while still others would buy fire protection from an adjacent town. Towns would also vary considerably on how much they relied on technology. Some might consider crime better controlled by electronics and others by policemen on the beat. Because the town is a small unit, the cost of pilot studies to test such alternatives would be low. Also, using the market creates a natural mechanism for the transmission of new technology into practice. Firms which did not keep up with technology would simply lose business. Moreover, the town manager could shop on the basis of price performance comparisons without having to keep abreast of the technology in all areas of town services. To assist in making such decisions, operational information policy would enable service firms to analyze alternatives and sell evaluation services.

While the metropolitan-town organization should lead to a more efficient delivery of local services satisfying the diverse needs of numerous lifestyles, giving the town the power to support the lifestyle of the majority would raise a fundamental issue of separation of church and state. Any lifestyle, whether it be secular family life, gay, or fundamentalist Christian, involves both questions of belief and the potential for conflict with rival beliefs. To reduce the conflict one might consider trying to expand the separation of church and state to a concept of separation of belief and state. This approach would create fundamental problems, because even science itself is based on belief. Science like religion cannot be based purely on reason, since, as David Hume demonstrated, experiments do not satisfy the assumptions for mathematical induction. From a general perspective, then, government cannot be separated from matters of belief. The resolution of this conflict is that governments higher than the lowest level would promote beliefs potentially refutable by experiments and would maintain separation of all beliefs based purely on faith. Beliefs based on faith would include traditional religions, ethical systems and any secular lifestyle. In promoting the lifestyle of the majority, then, towns could promote beliefs based on faith.

A town activity, which illustrates the problem raised by the need to separate nonrefutable beliefs and higher levels of government, is local education. Ideally for local education to be integrated into the community, primary and secondary education should be under the control of the towns. Advances in technology could will make this decentralization effective. But if local education is to be integrated into the community, local schools would be teaching nonrefutable beliefs as well as simply skills and this would cause a conflict with the separation of nonrefutable beliefs and the state under the condition that any portion of school funding comes from higher levels of government. How these problems can be resolved in a manner which also promotes an effective implementation strategy requires some discussion.

Two aspects of technological advance would enable very small schools to offer diverse, quality education. First, the advance in telecommunications would facilitate remote teaching, thus students would not have to be physically brought to a central location in large groups to achieve a diverse curriculum. Second, the development of educational software in the form of hypermedia, that is interactive voice, text, and image instruction material will enable the student to individually interact with his terminal. As technological forms of education advance, the teacher is freed to individually interact with students to help them select the most appropriate educational materials.

The decentralization scheme based on the assumption that a diverse, quality primary and secondary educational system can be created for a very small group of students, is presented with the caveat that numerous variations in this scheme are needed to empirically test alternative decentralization schemes. As with other multilevel government activities the role of the federal government would be promoting research and development. The role of the state governments would be focused on educational equity and the state research universities. To what extent should funding for education be equalized between poor and rich towns would remain a difficult problem for state governments. The state operational governance over education should be directed at the complex state research universities which would play an essential role in developing new technology and industries, training professionals, and educating the brightest undergraduates.

The metropolitan and district governments would have an expanded educational role. They would be responsible teaching colleges, junior colleges and adult education. The issue of primary and secondary educational performance standards would be decentralized to the metropolitan governments. The metropolitan government as part of its overall economic competitiveness strategy would thus be responsible for the educational quality of the local workforce.

The difficult question regarding decentralization is how much education should be decentralized from the metropolitan level to the town level. To obtain a better implementation strategy for innovation, the operational control of primary and secondary education would be delegated to the towns. This means that the choice of educational materials and curriculum would be determined by the towns. As higher educational performance can be achieved by having the most appropriate educational material for each individual student, selection of educational materials would become the task of the teacher, who would counsel the individual student to select the most appropriate material in the social nervous system. To ensure that all schools achieved at least minimal standards, most metropolitan governments would make the decentralization of administrative control over primary and secondary education conditional on the performance of students on standardized examinations.

Metropolitan governments, generally through public service corporations with appointed or less frequently elected trustees, would provide specialized services such as certain types of labs and other specialized equipment too expensive for most towns to purchase. Some students would go to specialized metropolitan schools and some would travel to specialized schools for short intervals to study a subject in a concentrated manner. Metropolitan teleconferencing would also be promoted for a variety of purposes such as specialized courses for which the demand at each town is insufficient such as special instruction of the gifted, disadvantaged, or students with special interests.

Decentralizing the operational responsibility for primary and secondary education to the towns, would enable the towns to integrate local education into the community lifestyle. In most towns the school system would be part of the local government and local school facilities would be used for multiple purposes day and night. Some towns would integrate before and after day care with their school systems and some would integrate the school athletic facilities into the overall athletic program for all town members. In general, schools would be used at night for adult education and other activities. As towns can promote systems of belief, towns could integrate any type of ethical of religious teaching with their secular education program. Because most states would have some educational funding equalization program in effect and some towns would be receiving research funds from the federal government, the problem of separation of nonrefutable beliefs and higher levels of government must be resolved. This problem should be handled by the concept of experimental precedents of the Constitution. In interpreting the constitution federal judges would establish a variety of precedents applicable in various groups of states. These precedents would vary from strict separation of nonrefutable belief activities from secular education to integration of activities provided the various components were appropriately funded. Over time empirical evidence would modify the precedents in keeping with the extent households had freedom of location. The greater households have freedom of location, the less need would exist for strict separation of activities.

Besides accommodating local integration of education and separation of nonrefutable beliefs and states, the proposed decentralization scheme would promote the separation strategy for innovation. The great advantage in decentralizing operational control of primary and secondary education to the town, then, would be to obtain much greater variation. Because the variation among the majority of the towns would be much greater than among the metropolitan regions, towns would naturally be prepared to implement a much wider range of alternatives, such as different technological approaches to education. With greater decentralization the federal government in funding new experiments would be more likely to achieve a better experimental design, given the much larger number of towns than metropolitan areas.

The decentralization of local education to smaller, more flexible units should also increase the rate of imitation. With federal government promotion of educational research, private firms and nonprofit organizations would use the advances in knowledge and educational methodology to create a wide variety of competing educational materials available through the social nervous system. Successful innovations would be rapidly imitated because of the creation of a larger number of smaller, more flexible education systems. Imitation can also be increased by having the metropolitan school districts compete for students in teleconferencing services.

Another empirically testable incentive mechanism to promote competition between alternative educational technology would be to fund primary and secondary education by vouchers allowing parents to select among alternative schools for their children. Parents would have four basic choices for educating their children: the local town school, a school at another physical location, a teleconferencing school through a terminal, or a home school organized by the parents using material available through the social nervous system. Assuming most parents would live in towns supporting their desired lifestyle, most parents would probably send their children to their town school system. But a system of vouchers increases competition because it would mean that a town school system could not automatically assume that most of the children in the town would attend the local school.

Individual

Regardless of how the political system is designed, it will not achieve good performance unless the individual voter demands good performance from his elected officials. As will be discussed, the promotion of close knit communities in the towns will encourage citizens to vote. Reorganizing the election process to provide the voters better information to evaluate the incumbent and his or her challengers should greatly reduce the cost of evaluation. Consequently voters will demand better performance.

As was pointed out in Chapter 5, towns were given broad powers to promote the lifestyle of the town majority in order to promote more congenial communities to compensate for the more impersonal, insecure market. In addition, to protect themselves from overzealous towns, individuals were given rights to limit the town's efforts to promote the majority lifestyle. As freedom of location increases, most individuals would seek a town which promotes their lifestyle. And as people's dependence on the automobile for work and shopping decreases, most towns would promote more personal interaction than currently.

The new design would promote individual participation in the political process. Because the new towns govern emotional lifestyle issues which directly affect them, citizens would want to vote either directly on such issues or for the candidates representing the various positions. Also, towns with similar political beliefs would have incentives to organize political action committees to encourage all citizens to vote in elections for higher levels of government. Because to the extent that the town tends to cast identical votes, the town's ability to influence the political process would be proportional to the extent that the town customarily turned out the vote. The advance in the social nervous system facilitates the formation of such special interests groups to promote political causes. For example, towns with similar political beliefs could easily organize into a league in order to obtain even more leverage on politicians at higher levels of government.

The participation of interested groups of citizens in higher levels of government, while desirable from the perspective of encouraging participation, would be subject to the same check of a professional review as any other special interest group. Indeed, the decentralization of lifestyle issues to the towns is designed to reduce, as much as possible, promotion of emotional issues at higher levels of government. The positive impact of the concerned citizens on promoting good government would not be a valiant attempt to promote some form of legislation which would not pass a professional review, but rather their impact would be to insist that politicians would have to be accountable for their performance. Since voters have limited cognitive skills and resources to evaluate their political alternatives, the information policy and election organization should be designed to improve the performance of voters under the conditions of modern government. Hence, the purpose of election information policy would be to make the voter aware of whether his government is providing competitive services. And as government has become so complex that having elections for all levels of government on the same day time blurs the focus of the voter, elections would be rearranged in order to simplify the choice process.

Accordingly, in order to focus voter and media attention on what services and how efficiently they are being provided, elections of each of the four levels of government would occur on the same day during an assigned quarter throughout the country. For example, town elections throughout the nation would occur on the same day in the winter quarter, although in direct-democracy towns, voters would vote on issues on a regular basis. Similarly, metropolitan elections throughout the nation would occur on the same day in the spring quarter, state elections on the same day in the summer quarter, and national elections on the same day in the fall quarter. This does not mean that every town, metropolis, or state would necessarily have elections every year, but on years in which they did they would have to hold them on the assigned day. Public campaigning for each election would be restricted to a few weeks before the respective election. Presidential elections would have a longer campaign period.

In addition, information policy would encourage more evaluation of political performance by comparisons with rival governments and less on personal discretions. Prior to each campaign period information policy would require the federal government to release of comparative figures for the various services at each level of government and the associated costs. At the national level this would require comparisons with other nations. The state and metropolitan governments would release additional information prior to the beginning of the campaign periods of subordinate government elections. The purpose of this release of information is to get the media focused on the comparison of costs and benefits of governments at the same level. For example, during the town election campaign national and local media coverage would be focused on innovative towns and high cost exceptions. Thus, news segments such as ABC's American Agenda would show during town elections, human interest stories of towns with successful innovations of better services at lower costs. Hopefully, only the tabloids would concentrate on the candidates' sexual peccadillos.

The release of such information would have a great impact on the political campaigns in informational society because of the greater political organization in informational society towns. Political action committees would digest election information and formulate the community position on election issues. Such political action committees would be aware of whether the incumbent's previous actions had resulted in negative professional reviews. But just important the political action committees would be aware of whether their government was providing quality services efficiently. Incumbents who did not rapidly imitate successful political innovations would be turned out of office by challengers promising to deliver to a more aware political audience. Thus political innovation and imitation would be accelerated.

Evaluation

The purpose of creating a professional review and decentralizing government was to increase the rate of innovation. As innovation is measured against the true but unknown common weal the first question to consider is whether the proposed design would improve the estimate of the common weal.

The criterion for a good estimate of the common weal is that it have the properties of general benefits, consistency and efficiency. As all levels of government would be subject to a professional review, such reviews would provide voters with easy to understand measures of political performance. But the extent that the professional review creates incentives for elected and appointed governmental officials depends on the political organization of towns. Under the proposed design, towns in promoting compatible lifestyles would tend to become relatively homogeneous and develop a strong sense of community. Hence, each town would organize politically in order to promote its common interests. Operational informational policy would make it much easier for politically active towns to monitor officials in higher levels of government. Because publicly funded elections would deny incumbents an overwhelming advantage over challengers, incumbents would have very strong incentives to avoid professional reviews.

Strong monitoring of politicians changes the nature of political factions trying to influence the political process through influence pedaling. As all elections would be publicly funded, factions would have few legal financial inducements for politicians. They would promote their interests by analytically demonstrating that their proposals had the properties of general benefits, consistency and efficiency while those of their opponents lacked the requisite properties. Thus an astute politician would use the rivalry of factions to ensure that his proposals would pass a professional review.

Greater decentralization would increase the property of general benefits because the proposed organization promotes homogeneity within political units and heterogeneity among political units. With homogeneous interests in subordinate governments the general benefits criterion is much easier to satisfy than the same criterion on legislation at the federal level. This is especially true at the level of the towns which were designed to promote lifestyle issues. To the extent that towns promote compatible lifestyles then promoting the lifestyle of the majority would have general benefits in most cases.

In as much as consistency has up to now not been considered a desirable property of government the imposition of a consistency criteria in the profession review would lead to more consistent estimates of the common weal than currently.

The driving force for efficiency in the decentralized government would be obtained more from competition between political units than the professional review. Because decentralization of governmental tasks places them in a more competitive political environment, efficiency considerations increase with decentralization. And the organization of elections and the prerelease of cost data on governmental services would make voters very concerned if their costs for government services were out of line with their competitors.

The proposed organization of a more decentralized government would improve the estimate of the common weal. This in itself would improve the rate of governmental innovation. The second aspect of improving innovation is improving the implementation strategy for innovation. In the proposed design the implementation strategy is made much more systematic than currently.

In the proposed strategy the federal government assumes the research and development strategy for innovation for all subordinate governments. The federal government through scientific information policy and grants for research promotes discoveries leading to governmental innovation. Where possible the federal government conducts experiments as the basis for innovation. For this governmental task the implementation strategy would become a separation strategy. However, as many aspects of social policy are not amenable to systematic experimentation, the empirical knowledge would be gained by the natural variation in objectives and policies of subordinate governments. The federal government would systematically analyze these variations and would offer incentives for pilot studies which explore new alternatives.

The proposed design would improve governmental innovation for three reasons. First, senators responsible for the federal research and development tasks would campaigning for reelection by extolling the merits of the innovations they had sponsored. In this regard the separation of federal research and development efforts from the operational control by subordinate governments creates a check on such senators. For a senator to claim an innovation, subordinate governments would have to adopt these potential innovations. Given the more competitive environments of subordinate governments, this would put a tremendous stress in innovation on cost effectiveness. But since Senators would have a six year reelection cycle, they would have a long enough planning horizon to accomplish significant improvements.

Second, the design and freedom of location encourages the heterogeneity between subordinate governments at each level. This would encourage diversity between governmental ends and means. Differences among ends and among means would be tested simultaneously rather than sequentially. Thus innovation would be accelerated because of the greater diversity between subordinate governments.

Third, the cost of mistakes would be decreased by greater decentralization. In selecting objectives and policies governments at all levels frequently make grave mistakes. Because of the proliferation of professional talent it is assumed that the states and metropolitan governments would have access to competent experts as well as the federal government. Thus it is assumed that the federal government would not have superior talent. Currently when the federal government makes a mistake, it is a colossal mistake. The greater diversity made possible by greater decentralization would mean that major mistakes would generally be localized to the subordinate government that made them.

Notes and References 9

1. Governments below the level of the state possess only those powers directly or indirectly granted to them by the respective state government. This delegation of power is known as Dillon's Rule after the judge ruling in *City of Clinton v. Cedar Rapids and Missouri Railroad Company* (1868)

2. Swiss, James E., 1984, Intergovernmental Program Delivery: Structuring Incentives for Efficiency, in Golembiewski, R. T. and A Wildavsky (eds), *The*

Costs of Federalism, (Transaction Books: New Brunswick)

3. Andrews, Richard N. L., 1984, Economics and Environmental Decisions, Past and Present, in Smith, Kerry V. (ed), Environmental Policy under Reagan's Executive Order, (University of North Carolina Press: Chapel Hill)

4. Boruch, R. F. and J.S. Cecil (ed), 1983, *Solutions to Ethical and Legal Problems in Social Research*, (Academic Press: New York)

5. The Advisory Commission on Intergovernmental Affairs, ACIR, has proposed five criteria for federal government decentralization: national purpose, economic efficiency, fiscal equity, political accountability, and administrative effectiveness. See: ACIR, 1981, An Agenda for American Federalism: Restoring Confidence and Competence, (A86), (US Government Printing Office: Washington). In this book no weight is placed on national purpose. The only consideration of equity is social inheritance and equal treatment before the law as modified by the need for experimental variation. Political accountability is considered separately in the general design of government and administrative effectiveness is considered part of economic efficiency.

6. An alternative method to achieve a more uniform decentralization would be for the smaller states to form compacts with the larger states. This approach will not be explored in this book.

7. Bowman, Ann O'M. and Richard C. Kearney, 1986, *The Resurgence of the States*, (Prentice-Hall: Englewood Cliffs)

8. See Chapter 7 of reference 7

9. For an argument supporting this position see: Hoenack, Stephen A., 1989, Group Behavior and Economic Growth, *Social Science Quarterly*, pp 744-758. This line of reasoning is obviously in conflict with Olson, Mancur, 1982, *The Rise and Decline of Nations*, (Yale University Press: New Haven). Freedom of location and the implementation of a professional review would tend to promote competitive, growth oriented government.

10. For a discussion of alternative approaches, see: Siebert, Horst, 1987, *Economics of the Environment, 2nd Edition*, (Springer-Verlag: NewYork)

11. Harrington, W. and A. J. Krunpnick, 1981, Stationary Source Pollution Policy and Choices for Reform, in Peskin, H. M., P.R. Portney and A. V. Kneese (eds), *Environmental Regulation and the U.S. Economy*, (John Hopkins University Press: Baltimore) 16. Among the first to recognize the virtues of the complex urban political economy were Bish and Ostrom. See: Bish, R. L. and V. Ostrom, 1973, *Understanding Urban Government*, (American Enterprise Institute: Washington)

12. For a discussion of worker safety issues, see: Chelius, James R., 1977, *Workplace Safety and Health: The Role of Workers' Compensation*, (American Enterprise Institute: Washington)

13. A good program of safety incentives can greatly reduce accidents and for some individual firms can be economically justified by the decrease in experience-rated workers' compensation premiums. See: Kendall, Richard M., 1986, Incentive Programs with a Competitive Edge, *Occupational Hazards*, March, pp 41-45. However, Chelius and Smith performed a statistical study

in which they did not observe an experience-rating effect on employer behavior. See: Chelius, James R and Robert S. Smith, 1983, Experience-Rating and Injury Prevention, in Worrall, John D.(ed), *Safety and the Work Force*, (ILR Press: Ithaca)

14. Indeed, if economic compensation is set too high, workers are encouraged to seek workers' compensation as a substitute for working. See: Worrall, John D.(ed), *Safety and the Work Force*, (ILR Press: Ithaca) One the other hand compensation should be commensurate with the magnitude of the loss of income and be adjusted for inflation.

15. For an extensive discussion of alternative arrangements see: Salvas, E.S., 1987, *Privatization: The Key to Better Government*, (Chatham House Publishers, Inc.: Chatham)

16. Four levels of government is assumed optimal for a bounded rational voter with limited resources. This assumption would be empirically tested by a wide variation in the political organization of local government in urban areas. Only the assumed ideal organization will be discussed.

17. In rural areas the large area-wide government that corresponded to the urban metropolitan government would be the district government. Below this level would be county governments for rural areas, and towns for incorporated areas. A large rural area unified by a shared interest, such as livestock production, would have a district government whose territory shifting in accordance with economic growth patterns. As most people live in urban areas the discussion in the text will consider only the metropolitan governments.

18. This approach has been proposed by Demsetz, Stigler and Posner in separate papers. See: Demsetz, H, 1968, Why Regulate Utilities?, *Journal of Law and Economics*, vol 11, pp 55-66; Stigler, George J., 1968, *The Organization of Industry*, (Richard D. Irwin, Inc: Homewood); and Posner, R.A., 1974, The Appropriate Scope of Regulation in the Cable Television Industry, *Bell Journal of Economics and Managements Science*, Vol 5, pp 335-358.

19. See Roth Gabriel, 1987, *Private Provision of Public Services*, (Oxford University Press: New York)

20. An extensive list is contained in reference 15.

Chapter 10

The Individual

Introduction

The discussion of the individual in informational society is reserved for last in order to describe the society in which the individual pursues his goal-directed activity. The design of informational society is expected to enable the individual, in pursuing his self interest, to promote the common weal. To achieve this result, individual self interest must be consistent with achieving a high rate of discovery, invention, and innovation. In informational society, this would be partially achieved through the numerous opportunities for adults to discover, invent, and innovate, activities which would be rewarded by such incentives as fame and fortune.

Intergenerational transfer

From the perspective of achieving a high rate of discovery, invention, and innovation, the development of individuals should, ideally, produce talented, well educated adults who actively seek better performance in all goal directed behavior. This ideal would be consistent with the interests of parents wishing to promote the success of their children in informational society.

It has always been true that over time the intergenerational transfer of genes, social values, and education is modified by social change. Medical discoveries such as insulin and better health care, for instance by enabling the population to manage what might formerly have been a deadly disease, in effect now promotes the transmission of genes for hereditary diseases such as diabetes[1]. Likewise, over time the Calvinistic work ethic has become secularized and increasingly hedonistic as leisure has increased. Parenting, too has been radically changed, with yet undetermined consequences, by the development of families in which both parents work outside the home. In response to these and other social developments the education of children is constantly being modified to correspond with changing political economic needs.

Improved social performance would result if changes in the intergenerational transfer were the result of a systematic strategy for innovation. Because of the personal nature of intergenerational transfer, decisions concerning this transfer have traditionally been decentralized to the level of the family with some governmental regulation, such as the requirement for sending children to school. Thus, intergenerational transfer is an excellent candidate for the use of separation strategy except for one problem. The great variation in towns in informational society should provide good nonexperimental data concerning intergenerational transfer decisions; however, because of the personal nature of many aspects of intergenerational transfer, the prospect of ever performing controlled experiments is extremely limited.

With advances in knowledge parents will leave the selection of their children's genes less to chance and more to direct manipulation. Already in current society the widespread diffusion of birth control technology has enabled parents to increase their control over the timing of children. In the future parents who desire a child of particular sex will increasingly use technology such as centrifuges to increase the probability of having a child with the desired sex. Currently, tens of thousands of women, primarily women whose husbands are infertile, have artificial inseminations. While society has performed selective breeding on plants and animals since recorded history, humans are reluctant to consider such practices in human reproduction; nevertheless, the trend in sperm banks for artificial insemination is to collect sperm from superior males who are free of known defects.

As discoveries in human genes are made, this knowledge will promote innovations in human reproduction decisions. An exciting project of the biologists, which will take decades or even much longer to complete, is mapping the genes in the human genome. While such knowledge, over the extremely long run, will enable parents to design their childrens' genes, this knowledge will create fundamental social problems which must be overcome. For example, genetic screening will stigmatise some individuals as unfit for marriage. Others for whom genetic screening indicates they are more susceptible to dangers in the workplace will have their employment options reduced. Also, individuals for whom genetic screening indicates they are susceptible to heart disease will have trouble obtaining insurance[2].

Hence, given the negative consequences of bad genes, parents will make ever effort to provide their offspring with the best genes possible. As discoveries of human genes will enable women to increasingly identify such genetic liabilities in their fetuses, they will have greater incentive to undergo amniocentesis in order to identify these disorders. More women may be faced with the difficult decision to abort, which will exacerbate the tensions between pro-choice and right-to-life factions. And if abortion becomes a common practice to avoid children with undesired characteristics whether genetic disorders, a particular sex or even eye color, the probable result will be an increase in the number of carriers of genetic disorders[3]. Consequently, as low cost vitro fertilization is developed in response to rising demand, couples with defective genes would have strong incentives to obtain an egg or sperm from a superior gene pool bank with the desired characteristics.

The long run promise of discoveries concerning human genes is bioengineering to create procedures for correcting genetic defects and increasing desirable characteristics such as human intelligence. The first success in gene therapy has been temporarily correcting for ADA deficiency by returning genetically corrected white blood cells to the body[4]. The long term solution will be to alter the genetic characteristics of these cells produced in the body. In the very distant, perhaps ever receding, future bioengineers will attempt to perform genetic surgery on the fertilized egg using in vitro fertilization. Parents who prefer their own genes to those of strangers would have strong incentives to use such procedures to avoid such disabling conditions such as Tay-Sach's disease, cystic fibrosis, hemophilia or Huntington's chorea. At first such procedures will be most in demand by couples wishing to avoid high probability of defective genes. But, in time couples may simply wish to use techniques which increase the chances of having not just a normal but a superior child.

Granted, human gene manipulation is and will no doubt continue to be a controversial area of technological advance. But in the proposed design much of the potential political conflict surrounding this issue would be prevented by the voluntary separation of people holding controversial beliefs or practicing controversial lifestyles into separate towns. In informational society the lifestyles of some towns would support the use of technology to improve the transfer of genes while others would be vehemently opposed. Although the proposed decentralization criterion would enable parents in any town to use gene technology, most parents wishing to use such technology would probably choose to reside in towns supporting such practices. To the extent that such practices demonstrated improved performance, they would gradually become widely practiced and perhaps would eventually even become acceptable among segments of the population which had previously opposed them.

The second component of the intergenerational transfer are the social values a young person acquires during maturation. Most innovation in parenting would be focused at improving the performance of specific problems such as innovations to improve the parenting performance of working parents. This is one area in which government would be willing to subsidize experiments in alternatives systems of parental leave, child care, and the integration of parenting and work. Government would also take an active role in experiments directed toward reducing social problems such as teenage gangs and the intergenerational transfer of child abuse.

The evolution of values in informational society would stem primarily from the natural evolution of values currently held by industrial society. In informational as in current society, financial rewards and status in political economic activities would be based on the success an individual achieved in his or her chosen career. Communities of successful people would strongly value, either explicitly or implicitly, effort put forth to perform well in goal-directed activities. Such efforts are consistent with a wide variety of lifestyles, such as those based on religious as well as secular variations of the work ethic. Such striving is also consistent with modern psychological concepts such as self-actualization. Thus it is expected that parents in success-oriented towns would transmit to their children the value of achievement through effort and the willingness to assume moderate risks.

As society becomes increasingly leisure-oriented, however, the work ethic and the value system supporting it could disappear. With more leisure time, most individuals will focus their lives more towards leisure alternatives than they do currently. This move to a leisure-oriented society could lead to the development of a value system in which the good life means an absence of striving towards goals which require the development of human skills. For example, a society which places a great deal of value on leisure activity may move towards a completely hedonistic lifestyle. Or, alternatively, a very passive value system might evolve if leisure acquires the meaning of purely passive activities such as watching television. Under these circumstances few children are likely to grow up wishing to vigorously pursue discovery, invention and innovation.

The key to the evolution of successful value systems for informational society, then, is the actual development of leisure. As the time spent in work decreases, challenging leisure activities involving striving for performance in a goal-directed activity increase. One example would be the development of participatory sports for all ages, with competition promoted by matching teams of approximately equal ability. Another example would be the development of artisan activities by means of shows and competitions that promote talent. If leisure were developed in this manner, then value systems stressing the importance of striving for performance in goal-directed activities would be consistent with both work and leisure.

The need to promote goal directed leisure would lead to the innovation of government as a promoter of active leisure. Consider, for example, athletics. Participating in an active sport on a regular basis is much better for one's health than passively watching professional sports on television. And because good health has positive social benefits beyond the reduction of health costs to the individual, government has some incentive to promote active sports as a public good. There are, of course, many very different approaches to promoting such a policy. On the one hand, the government could subsidize sports facilities, or on the other, it might favor individual incentives, such as allowing insurance costs to reflect physical fitness. Similarly, government will promote other types of goal-directed leisure. In this way the goal of transmitting social values that promote informational society will be realized.

Another aspect of the maturation process shaping adult individuals is primary and secondary education. The purpose of educating children in informational society is the same as in previous societies–namely, to prepare children to lead constructive lives as adults. The emphasis of primary education is to teach basic skills and of secondary education to provide the basic foundations for making adult choices in political economic roles.

As was discussed in the previous chapter, the design of the primary and secondary educational system would place greater emphasis on technological innovation than the current educational system does. The focus of technological innovation would create methods of instruction combining teachers and technology which would respond to individual differences. Consequently, each student should receive a better education because innovations in educational technology would be to create an educational system which better responds to the educational needs of each student. Moreover, the development of educational technology should make for greater equality of educational opportunity, since the low cost of replicating educational software would make such technology available to all school systems.

To prepare students for life as adults in informational society will require an important shift in the focus of primary and secondary education. Adults in informational society will increasingly face a sequence of one time situations. For example, the purchase and implementation of new technology in work or the decision to purchase a new type of technological product as a consumer. To prepare adults for a future in which they must solve a sequence of non recurring problems, the education of children must place a much greater emphasis on problem solving and less on rote. Throughout their lives students will be facing constant learning situations where they must learn new skills to be employable. Faced with a new situation, students need to know how to find reference material in the social nervous system, how to learn the material using technological forms of education, and how to use software packages to use the new knowledge.

The Problem of Choice

Adults in informational society, as in previous societies, will face difficult choice problems in making such major decisions as the choice of a mate, career, or residence. In particular, adults in informational society will face greater difficulty than adults in previous societies in making secondary sequential decisions such as consumption decisions. First, an increase in the rate of discovery, invention, and innovation will increasingly create a unique sequence of decisions for individuals. For example, individuals making consumption decisions will find their tasks becoming less repetitious as the consumer faces a stream of new products to evaluate. A sequence of unique decisions to make will further become the norm as the routine aspects of work are increasingly automated.

Besides facing a sequence of increasingly unique choices, decision makers will generally have a much wider selection of choices for each decision. One factor in increasing the number of choices is that automation will reduce the cost of batch production, and this will enable producers to efficiently produce a wider assortment of goods. Also with activities shifting to the social nervous system, individuals will have a larger number of alternatives to consider.

The problem of choice has already been discussed with respect to the consumer and producer. To consider the issue further, let us consider the problem of adult education. Earning an income in informational society will require individuals to make a sequence of educational decisions. As was discussed in the chapter on economics, most jobs will last for only a finite time, to accommodate the constant technological change. Individuals must therefore develop strategies to cope with constant employment changes, since most will go through a life cycle of work combined with training for the next job. Some, in more stable industries, will mesh the upgrading of their skills with work to maintain a fairly stable work experience. Others, who are displaced by technology, will have to train for jobs in completely new industries. To compete for employment opportunities, then, most individuals will have to constantly learn new knowledge and procedures all their working lives.

The advance of technology will provide adults with much greater choice in educational alternatives than exists currently. With advances in computer software, one alternative will be self-paced instruction through a terminal and another will be courses organized through teleconferencing. With the resulting freedom of location, students would by able to obtain education through the social nervous system from a very large number of suppliers such as universities, colleges, technical schools and corporate training programs. Another factor which would give students many more alternatives is the flexibility inherent in technological education. The student can interact with the software at his discretion, hence the student would have considerable control over how he integrated education into his daily life. Students would need to attend classes at specific locations only if there were specific reasons for doing so, such as the need to manipulate physical objects, as is the case in many vocational training, art, or music courses. But, the need to be in a specific location will decline with advances in remote control, virtual reality and teleconferencing.

The choice of travel to obtain post secondary education in informational society would increasingly become an option and not a necessity. Many young adults would be able to select educational options that would minimize travel and living costs. They might, for instance, want to reside with their parents until they could afford their own residences. Given the transfer of education to the social nervous system, such individuals could obtain quality education. Once established, most adults would select educational options that did not involve much travel in mixing education with their careers.

Some young adults, of course, would have strong incentives to combine travel with their education. With the growth of multinationals and other international institutions, an ambitious individual would have incentives to develop social skills in dealing with individuals from many cultures. Because many of the courses would be taught through the social nervous system, a student would be free to develop these skills by actually traveling and while simultaneously pursuing his education. For example, a youth could travel around the world on a grand tour and at the same time use rented terminals or a portable terminal to continue his formal studies. To accommodate such cases, universities would establish worldwide networks enabling students to transfer between campuses in different countries. Accordingly, students might relocate once a semester on one of campuses in the network. The student would take some courses live at the school of residence and pursue the rest of his studies through the social nervous system. This development would be an evolution of the current year-abroad programs at many universities. This type of educational opportunity would also appeal to middle level executives from multinational corporations considering mid-career training.

The sequence of adult educational choices is typical of the difficult choice problems facing individuals in informational society. To improve their performance in decision making individuals will increasingly make use of decision support systems available through the social nervous system for all types of decisions. The development of such systems will be an extension of the type of decision support systems used currently in making business decisions. How such systems might arise for consumer decisions was previously discussed. Also discussed previously was the evolution of such systems for voter issues and for the choice of a residence in a town supporting the desired lifestyle.

Because a large number of individuals in informational society will face a sequence of job and education decisions, entrepreneurs will create numerous decision support systems to aid in making job and education decisions. Besides listing all the job and education opportunities, such systems would offer numerous services. One service, for instance, might be to screen all the possibilities to determine the alternatives open to the individual. Another service would be aptitude and preference tests to help counsel individuals into making better decisions.

Decision support system are also likely to be created for the selection of leisure pursuits. For most individuals in informational society leisure would become an avocation which adds to the quality of life, and in keeping with this, many leisure activities would be the kind that would challenge human abilities. A continual supply of new leisure opportunities of this sort would be created by advances in technology . Examples of technological advances which create new leisure in current society are the invention of hand gliders and computer games.

Individuals would also make increasing use of choice support systems in making personal choices such as finding a date or a mate. Currently young professionals use computerized dating to decrease the search cost of finding compatible dates and potential mates. The widespread acceptance of such information services for personal choices, however, depends on the long-term performance of such services. With the divorce rate currently standing at fifty percent, it seems that the systematic application of discovery to computerized matching could hardly do worse than the system of intuitive matching of today. Scientific information rights would create the basis for gradual improvements in matching algorithms for these types of choices. If evidence accumulates that the use of software to aid in the decision making involved in mate selection leads to more satisfying relationships and to more successful marriages, then such practices will grow.

The more difficult issue in the development of decision support systems is the conflict between efficiency and privacy. If a date-matching service developed a large number of clients, then programs would be used for preliminary screening to find possible matches. The performance of these programs would depend on access to data most individuals would probably not want to have casually released for perusal. The compromise would be to have one data base of raw data for perusal by prospective clients and a much larger data base for the matching programs. To obtain the larger database voluntarily, the matching service would have to develop a carefully maintained reputation for discretion and deliver much better results than intuition.

The move to use choice support systems in all types of choice making should improve the quality of choices for a variety of reasons. First, decision aids will reduce the deficiencies in behavioral man's reasoning. Second, with software support, individuals will gain sophisticated statistical reasoning procedures much better than the limited-incorrect-intuitive statistical procedures of bounded rational humans. Third, in constructing objectives choice aids will generate greater consistency among the selection of alternatives. Fourth, with operational information policy, data files are more likely to contain both the positive and negative aspects of the alternatives in a choice–a point, which was discussed in great detail in the creation of information services for consumers. Fifth, scientific information policy provides the basis for researchers to constantly create better evaluation tools such as simulation software allowing the user to analytically study alternatives. And finally, making choices in a computer environment allows the user to inexpensively evaluate large numbers of alternatives.

In summary, decision supports systems will improve the goal directed performance of individuals in two important respects. First, these systems will enable individuals to make better estimates of their goals and secondly, they will increase the ability of individuals to select the best alternative. In this respect the performance of behavioral man should more closely approximate the ideal of rational man.

To some readers the move to using computerized decision support systems to aid in choices of even a very personal nature would be an indication that informational society would be much less personal than current society. But in fact, the move to decision support systems should help to create a more personal society. To be sure the move to decision support systems would greatly reduce human role relationships such as between a sales clerk and a customer. But, by reducing the time required in this sort of role relationships decision support systems could create more time for personal relationships in the community supporting the individual's chosen lifestyle. The development of personal relationships would be further enhanced by some of the greater personal freedoms an individual would have in informational society to choose. First of all a fundamental choice of every adult is the town in which he chooses to reside. With freedom of location, almost every individual should be able to find a town which supports his lifestyle. And as individuals will have greater leisure they will have more time to develop personal relationships. Because most individuals would not have to relocate to pursue their careers, they would have the time to develop long term personal relationships.

It is important to remember, moreover, that humans have the capacity to humanize the use of technology. Shopping through the social nervous system does not necessarily mean a lonely individual glued to his computer screen. Socially oriented people are currently making shopping through telemarketing a social event. In the future such people would make imaginative use of visual simulations, such as trying on unusual clothes or simulating new furniture in a room for humor as well as for purposeful shopping.

Individual and Social Performance

Social performance in informational society would be as much a product of individual performance as is the case now. The complex relationships between science, markets, government and the individual in informational society require a complex incentive structure to harness individual effort to the common weal. To examine this issue in greater detail, let us consider innovation. Throughout this book it has been advocated that achieving a higher rate of innovation could be achieved by shifting to a better strategy for implementing innovations. But to achieve a this higher rate of innovation also depends on individual performance within a better institutional arrangement for innovation.

To discuss the role of individual performance in achieving a higher rate of innovation let us first review the principle features of the proposed separation strategy. All innovation would be based on a systematic approach to discovery which promotes innovations. For this purpose the federal government would fund basic research, and scientific information policy would provide systematic observations. In addition, for some types of innovation, the government would conduct experiments to systematically test alternatives. Such experimentation will be necessary, as for the foreseeable future the social sciences are unlikely to be sufficiently developed to implement innovations without extensive empirical tests. Finally, systematic analysis of attempts at innovation will promote faster imitation.

One area where innovation would remain less systematic is the experimental design for testing alternatives. For some activities–for example, complete automated plants-the cost of testing alternatives is too great to allow systematic testing. Still other activities, both political and personal are too risky or too controversial to gain public support for systematic testing of alternatives. For a great many types of innovation, then, the testing of alternatives must be left to those who actually participate in the innovative activity. When such individuals or groups believe the implementation of the new alternative would improve performance in a goal-directed activity, they will test new alternatives to determine if this is in fact the case.

As was pointed out for these situations, where the decision maker bears the risk and the cost of implementing a new alternative, the experimental design can be improved by increasing the number of participants and by using subsidies. For political activities, improvements can be made by greater decentralization and subsidies that encourage risky alternatives with great promise. Better business innovation can be obtained by promoting consortia to promote applied innovation research and promoting startups to achieve a greater number of actual implementations.

To achieve a high rate of innovation, individual performance is required in the various innovation tasks, which as was pointed out in Chapter 1 included identifying evaluating, and selecting alternatives. The design of informational society would provide incentives for individuals in the various tasks creating innovations. Scientists making discoveries promoting innovations would win fame and financial rewards; business entrepreneurs would have financial incentives; and political entrepreneurs would have the political incentive of exercising power. The proposed community structure would promote social and personal experimentation because some towns would attract risk loving individuals who would support experimentation. And these lifestyle entrepreneurs would have the admiration of imitators. Provided that intergenerational transfer in informational society produces a sufficient number of striving individuals prepared to assume the moderate risks of innovation, informational society will have a high rate of innovation.

Notes and References 10

1. Novitski, Edward, 1977, *Human Genetics*, (Macmillan Publishing Co.,: New York)

2. Weinberg, R. A., 1991, The Dark Side of The Genome, *Technology Review*, Apr, pp 45-51

3. See reference 1.

4. Erickson, D, 1992, Genes to Order: Companies queue up to realize the promise of gene therapy, *Scientific American*, June, pp 112-113

Index